Selective Trial Test Package

Set 3
Second Edition

Sibashis Nanda
Founder & CEO, Learnikx Education

Student Name : _____

Learnikx Education
https://learnikx.com.au

Selective Trial Test Package Set 3
Learnikx Education

Copyright © 2025 Learnikx Education
https://learnikx.com.au

All rights reserved.

No part of this book may be reproduced, stored in a retrieval system or transmitted in any form or by any means, electronic, mechanical, photocopying, recording, scanning or otherwise, without either the prior written permission of Learnikx Education (learnikx.com.au), or authorization through payment of appropriate per-copy fee. No patent liability is assumed with respect to the use of the information contained herein. Although every precaution has been taken in the preparation of this book, Learnikx Education assume no responsibility for errors or omissions, nor is any liability assumed for damages resulting from the use of the information contained herein.

ISBN **978-1-922608-38-3**

Trademarks
All terms mentioned in this book that are known to be trademarks or service marks have been appropriately capitalized. Learnikx Education cannot attest to the accuracy of this information. Use of a term in this book should not be regarded as affecting the validity of any trademark or servicemark.

Warning and Disclaimer
[1] Every effort has been made to make this book as complete and as accurate as possible, but no warranty or fitness is implied. The information provided is on an "as is" basis. Learnikx Education and its owners shall have neither liability nor responsibility to any person or entity with respect to any loss or damages arising from the information contained in this book.
[2] This is not an officially endorsed training or publication of the NSW Selective Test. This book is developed independently of Australian governments or other agencies.

Contents

	Page
Series I Reading Comprehension Trial Test	1
Series I Mathematical Reasoning Trial Test	10
Series I Thinking Skill Trial Test	24
Series J Reading Comprehension Trial Test	37
Series J Mathematical Reasoning Trial Test	47
Series J Thinking Skill Trial Test	59
Series K Reading Comprehension Trial Test	71
Series K Mathematical Reasoning Trial Test	81
Series K Thinking Skill Trial Test	94
Series L Reading Comprehension Trial Test	106
Series L Mathematical Reasoning Trial Test	116
Series L Thinking Skill Trial Test	129
Series I Answers and Solutions	142
Series J Answers and Solutions	191
Series K Answers and Solutions	243
Series L Answers and Solutions	288

We Want to Hear from You!

Welcome to the second edition of *Selective Trial Test Package Set 3!* As the reader of this book, you are our most important critic and commentator. We value your opinion and want to know what we're doing right, what we could do better, what areas you'd like to see us publish in, and any other words of wisdom you're willing to pass our way. You can email and let us know what you did or didn't like about this book, as well as what we can do to make our books stronger.

Please note that we cannot help you with technical problems related to the topic of this book, and we might not be able to reply to every message.

When you write, please be sure to include this book's title and author as well as your name and contact information.

Email: admin@learnikx.com.au

Preptive Prepare Better
Series I Reading Comprehension Trial Test

Name: _____

Date: _____

Direction for Questions: 1 to 5

Read the two extracts and then answer the questions.

Extract A: The Art and Science of Content Creation

In today's digital age, content creation has become both an art form and a science. From captivating blog posts to engaging videos and immersive social media campaigns, content creators wield a diverse set of tools and techniques to connect with audiences and convey their message effectively.

At the heart of content creation lies storytelling. Whether it's through words, images, or multimedia, compelling narratives have the power to captivate audiences and leave a lasting impression. By understanding their target audience and crafting stories that resonate with their interests, content creators can foster genuine connections and build loyal communities around their content.

However, storytelling alone is not enough. Successful content creation also requires a strategic approach informed by data and analytics. By leveraging insights into audience behavior, preferences, and trends, creators can tailor their content to maximize engagement and reach. From keyword research to A/B testing and performance tracking, data-driven decision-making enables creators to refine their content strategy and optimize their impact.

In addition to creativity and strategy, adaptability is essential in the ever-evolving landscape of content creation. Platforms and algorithms change, trends come and go, and audience preferences shift over time. Content creators must stay agile and proactive, experimenting with new formats, technologies, and distribution channels to stay relevant and competitive in the digital space.

Collaboration also plays a vital role in content creation. Whether working with fellow creators, brands, or industry experts, partnerships can enhance the quality and reach of content, leveraging shared expertise and resources to create more impactful campaigns. By fostering a spirit of collaboration and mutual support, creators can unlock new opportunities and expand their reach beyond their individual capabilities.

Ultimately, content creation is both an art and a science—a delicate balance of creativity, strategy, adaptability, and collaboration. By mastering these elements and staying true to their unique voice and vision, content creators can inspire, educate, and entertain audiences around the world, leaving a lasting legacy in the digital realm.

Extract B: The Power of Authenticity in Content Creation

In a digital age dominated by curated feeds, sponsored posts, and influencer culture, authenticity has emerged as a powerful currency in the realm of content creation. Authenticity is more than just a buzzword; it's a guiding principle that shapes the way creators connect with their audience, build trust, and foster meaningful relationships in a crowded and competitive online landscape.

At its core, authenticity in content creation is about being genuine, transparent, and true to oneself. It's about embracing one's unique voice, experiences, and perspective, rather than conforming to trends or trying to emulate others. Authentic content reflects the true essence of the creator, resonating with audiences on a deeper, more emotional level.

One of the key elements of authenticity is transparency. Creators who are transparent about their motivations, affiliations, and processes build trust with their audience. Whether it's disclosing sponsored content, acknowledging mistakes, or sharing behind-the-scenes insights, transparency fosters a sense of openness and honesty that strengthens the bond between creator and audience.

Moreover, authenticity entails staying true to one's values and principles. Creators who align their content with their beliefs demonstrate integrity and authenticity by standing by what they believe in, even in the face of criticism or controversy. By remaining steadfast in their convictions, these creators attract like-minded individuals who resonate with their values, fostering a loyal and engaged community.

Authentic content creation also involves actively listening to and engaging with the audience. Creators who genuinely care about their audience's feedback, questions, and concerns demonstrate a willingness to listen and respond, fostering a sense of connection and community. By engaging authentically with their audience, creators cultivate meaningful relationships based on mutual respect and understanding.

One of the most compelling aspects of authentic content is its ability to create genuine connections between creators and their audience. When creators share their personal stories, struggles, and triumphs authentically, they invite their audience into their world, forging a sense of intimacy and camaraderie. This sense of connection fosters loyalty and engagement, as audiences feel seen, heard, and understood by the creators they follow.

Moreover, authenticity in content creation breeds authenticity in engagement. When creators respond to comments, messages, and feedback authentically, they demonstrate a genuine interest in their audience's thoughts and opinions. This two-way dialogue fosters a sense of community and belonging, as audiences feel valued and appreciated by the creators they support.

In addition to fostering genuine connections with their audience, authenticity also sets creators apart in a crowded and competitive landscape. With so much content vying for attention, audiences are drawn to creators who stand out by being genuine, relatable, and authentic. Creators who embrace their unique voice and perspective cut through the noise, capturing the attention and loyalty of their audience.

Furthermore, authenticity in content creation builds credibility and trust. When creators are transparent about their motivations, affiliations, and processes, they earn the trust of their audience. This trust forms the foundation of a strong and enduring relationship, as audiences feel confident in the integrity and authenticity of the content they consume. Ultimately, authenticity in content creation is not just a trend or a marketing strategy—it's a fundamental principle that shapes the way creators connect with their audience, build trust, and foster meaningful relationships online.

Question: 1 of 30

What is the central theme of Extract A?

- A. The importance of authenticity in content creation
- B. The role of storytelling in digital content
- C. The significance of collaboration among content creators
- D. The impact of data-driven decision-making in content strategy

Question: 2 of 30

According to Extract A, what essential role does adaptability play in content creation?

- A. It limits the scope of content diversity
- B. It ensures consistent adherence to trends
- C. It enables creators to navigate evolving platforms and algorithms
- D. It hampers collaboration opportunities with other creators

Question: 3 of 30

What role does audience engagement play in authentic content creation, as described in Extract B?

- A. It is unnecessary for maintaining authenticity
- B. It serves as a means to manipulate audience perceptions
- C. It demonstrates a genuine interest in the audience's feedback
- D. It creates barriers between creators and their audience

Question: 4 of 30

How does authentic content creation contribute to setting creators apart in a competitive landscape, as discussed in Extract B?

- A. By conforming to industry trends and standards
- B. By producing content solely for financial gain
- C. By embracing a unique voice and perspective
- D. By imitating the content of other successful creators

Question: 5 of 30

In Extract A, what is emphasized as a fundamental aspect of successful content creation?

- A. The utilization of advanced navigation technologies
- B. The reliance on traditional storytelling methods
- C. The importance of adapting to evolving platforms
- D. The integration of real-time audience feedback

Direction for Questions: 6 to 12

Read the two extracts and then answer the questions.

Extract A: The Enigmatic Beauty of the Red Sea

Nestled between the African and Arabian continents lies a marine marvel of unparalleled beauty: the Red Sea. Renowned for its crystal-clear waters, vibrant coral reefs, and diverse marine life, the Red Sea beckons explorers and nature enthusiasts alike to uncover its hidden treasures.

One of the most captivating features of the Red Sea is its rich biodiversity. The warm, nutrient-rich waters provide an ideal habitat for a dazzling array of marine species, from colorful reef fish to majestic sea turtles and elusive sharks. Coral reefs, teeming with life, adorn the seabed, forming intricate ecosystems that support countless organisms. These reefs are not only a sight to behold but also play a crucial role in maintaining the health of the ocean.

The Red Sea's unique geological history adds to its allure. Surrounded by arid landscapes and ancient civilizations, the sea has witnessed the rise and fall of empires, leaving behind a tapestry of cultural heritage. From ancient trade routes to sunken shipwrecks, the seabed holds clues to centuries of human history, enticing archaeologists and historians to explore its depths.

But perhaps the most enchanting aspect of the Red Sea is its sense of tranquility and serenity. Away from the hustle and bustle of modern life, the sea offers a sanctuary for those seeking solace and connection with nature. Snorkelers and divers glide through the water, mesmerized by the kaleidoscope of colors and the gentle sway of coral gardens. Sunset cruises offer a front-row seat to nature's daily spectacle, as the sky transforms into a palette of fiery hues over the tranquil waters.

Yet, amidst its beauty, the Red Sea faces threats from human activities. Overfishing, pollution, and coastal development pose risks to its delicate ecosystems, endangering the very wonders that make it so extraordinary. Climate change further compounds these challenges, with rising sea temperatures and ocean acidification threatening coral reefs and marine life.

Despite these challenges, conservation efforts are underway to protect the Red Sea and its inhabitants. Marine protected areas, sustainable fishing practices, and community-based initiatives aim to preserve the sea's natural beauty and promote responsible tourism. Through education and awareness, we can ensure that future generations continue to marvel at the enigmatic beauty of the Red Sea and its irreplaceable wonders.

Extract B: Guardians of the Red Sea

Nestled between the African and Arabian continents lies a marine wonderland of unparalleled beauty and biodiversity: the Red Sea. This vibrant ecosystem, teeming with life, is not only a natural treasure but also a fragile ecosystem in need of protection. Fortunately, a dedicated group of individuals stands as the guardians of the Red Sea, working tirelessly to ensure its preservation for future generations.

At the forefront of conservation efforts are marine biologists and scientists who devote their expertise to understanding and monitoring the Red Sea's delicate ecosystems. Through research expeditions, biodiversity surveys, and monitoring programs, these guardians gather valuable data on the health of coral reefs, marine life, and water quality. Their insights inform conservation strategies and management plans aimed at safeguarding the Red Sea's biodiversity.

Local communities also play a crucial role in protecting the Red Sea. Fishermen, tour operators, and coastal residents rely on the sea for their livelihoods, making them important stakeholders in conservation efforts. Community-based initiatives empower these individuals to adopt sustainable practices that minimize their impact on marine resources. By promoting responsible fishing, waste management, and tourism practices, these guardians help preserve the Red Sea's natural beauty and support the livelihoods of coastal communities.

Government agencies and policymakers are instrumental in enacting laws and regulations to protect the Red Sea's marine environment. From establishing marine protected areas to enforcing fishing quotas and pollution control measures, these guardians work to ensure the sustainable management of marine resources. By collaborating with stakeholders and enforcing regulations, they help mitigate threats to the Red Sea's ecosystems and promote long-term conservation efforts.

Non-governmental organizations (NGOs) and conservation groups also play a vital role in protecting the Red Sea. These organizations leverage their resources and expertise to implement conservation projects, raise awareness, and advocate for the protection of marine habitats. Whether conducting outreach programs, organizing beach cleanups, or lobbying for policy change, these guardians are tireless advocates for the Red Sea's preservation.

Tourism operators have a responsibility to promote sustainable tourism practices that minimize negative impacts on the Red Sea's ecosystems. By educating visitors about the importance of conservation and encouraging responsible behavior, these guardians help ensure that tourism contributes to the protection of the marine environment rather than its degradation.

Together, these guardians of the Red Sea form a formidable alliance dedicated to preserving one of the world's most precious marine ecosystems. Their collective efforts underscore the importance of collaboration, stewardship, and collective action in protecting our oceans for future generations. As long as these guardians remain committed to their cause, the Red Sea will continue to thrive as a beacon of biodiversity and natural beauty for generations to come.

Question: 6 of 30

What role do marine biologists and scientists play in protecting the Red Sea, according to Extract B?

- A. Advocating for sustainable tourism practices
- B. Monitoring and researching the Red Sea's ecosystems
- C. Enforcing fishing quotas and pollution control measures
- D. Organizing outreach programs for local communities

Question: 7 of 30

According to Extract A, what threatens the beauty of the Red Sea's delicate ecosystems?

- A. Overfishing and pollution
- B. Climate change and rising sea temperatures
- C. Coastal development and tourism
- D. Industrial activities and shipping

Question: 8 of 30

What is the primary message conveyed in Extract B regarding sustainable tourism?

- A. Tourism operators should prioritize profit over environmental concerns.
- B. Tourists should avoid visiting the Red Sea to minimize their impact.
- C. Tourism can contribute positively to conservation efforts when managed responsibly.
- D. Sustainable tourism practices have no significant impact on the Red Sea's ecosystems.

Question: 9 of 30

Which aspect of the Red Sea's beauty is emphasized in Extract A?

- A. Its role as a sanctuary for marine life
- B. Its tranquility and serenity away from modern life
- C. Its vibrant coral reefs and diverse marine species
- D. Its geological history and ancient civilizations

Question: 10 of 30

What distinguishes the Red Sea as a marine marvel, according to Extract A?

- A. Its proximity to African and Arabian continents
- B. Its vulnerability to human activities
- C. Its unique geological history and biodiversity
- D. Its popularity as a tourist destination

Question: 11 of 30

How does Extract B characterize the role of tourism operators in protecting the Red Sea's ecosystems?

- A. Prioritizing environmental concerns over profit
- B. Avoiding tourism activities altogether
- C. Promoting sustainable tourism practices
- D. Disregarding the impact of tourism on the environment

Question: 12 of 30

According to Extract A, what contributes to the allure of the Red Sea?

- A. Its vulnerability to climate change
- B. Its popularity as a tourist destination
- C. Its tranquil and serene environment
- D. Its proximity to African and Arabian continents

Direction for Questions: 13 to 17

Read the poem and then answer the questions.

How Sleep the Brave
by William Collins

How sleep the brave, who sink to rest
By all their country's wishes blest!
When Spring, with dewy fingers cold,
Returns to deck their hallow'd mould,
She there shall dress a sweeter sod
Than Fancy's feet have ever trod.
By fairy hands their knell is rung;
By forms unseen their dirge is sung;
There Honour comes, a pilgrim grey,
To bless the turf that wraps their clay;
And Freedom shall awhile repair
To dwell, a weeping hermit, there!

Question: 13 of 30

What is the implied meaning of the phrase "How sleep the brave"?

- A. The bravery of soldiers during sleep
- B. The peaceful rest of courageous individuals
- C. The lack of bravery in sleeping heroes
- D. The bravery of those who never sleep

Question: 14 of 30

Which natural element is personified in the poem?

- A. Wind
- B. Rain
- C. Spring
- D. Sun

Question: 15 of 30

What does the line "She there shall dress a sweeter sod" suggest?

- A. The beauty of a sunny day in spring
- B. The transformation of the burial ground by nature
- C. The arrival of fairies to adorn the graves
- D. The mourning of unseen forms

Question: 16 of 30

What is the role of Honour in the poem?

- A. Singing a dirge for the fallen soldiers
- B. Blessing the clay that covers the heroes' graves
- C. Personifying the beauty of nature
- D. Decorating the graves with flowers

Question: 17 of 30

Which emotion is attributed to Freedom in the final stanza?

- A. Joy
- B. Sorrow
- C. Anger
- D. Apathy

Direction for Questions: 18 to 24

Seven sentences have been removed from the text. Choose from the sentences (A – H) the one which fits each gap (1 – 7). There is one extra sentence which you do not need to use.

The Industrial Revolution: A Turning Point in Human History

1..... This period transformed society in profound ways, leading to significant changes in the economy, technology, politics, and culture.

Before the Industrial Revolution, most people lived in rural areas and worked in agriculture. 2...However, the invention of new technologies such as the steam engine and the power loom led to a shift from manual labor to machine-based production. This, in turn, resulted in increased productivity and economic growth.

With the rise of new technologies, factories emerged as the primary centers of production. Factories brought together large numbers of workers under one roof, allowing for efficient and standardized production of goods. 3....

The Industrial Revolution also led to a rapid increase in urbanization. 4.... This rapid urbanization led to the growth of slums, overcrowding, and poor living conditions. However, it also led to the development of new social movements and political reforms, as workers organized to fight for better working conditions and wages.

The Industrial Revolution was a period of immense technological advancement. Inventions such as the steam engine, the telegraph, the telephone, and the electric light bulb revolutionized transportation, communication, and everyday life. 5...

The Industrial Revolution did not only impact the countries where it originated. Its effects were felt around the world, as industrialized nations sought raw materials and markets for their manufactured goods. 6....

The Industrial Revolution remains one of the most significant events in human history. Its impact is still felt today in various ways, including, Economic Growth, Standardized Production, Urbanization, Technological Advancements, Global Interconnectedness.
While the Industrial Revolution brought about significant benefits, it also had negative consequences, such as environmental pollution, social inequality, and the exploitation of resources and labor. 7.....

- A. The economy was primarily based on manual labor, and production levels were relatively low.
- B. This led to the development of mass production, which made goods more affordable and accessible to a wider range of people.
- C. These technological advances laid the foundation for the modern world we live in today.
- D. This led to the expansion of colonialism and the exploitation of resources in many parts of the world.
- E. People moved from rural areas to cities in search of jobs and opportunities
- F. The Industrial Revolution has both positive and negative impacts.
- G. The Industrial Revolution was a period of rapid industrialization that began in Great Britain in the 18th century and spread to other parts of the world in the 19th and 20th centuries.
- H. As we face the challenges of the 21st century, it is important to learn from the history of the Industrial

Revolution and strive to build a more sustainable and equitable future for all.

Question: 18 of 30

Which one fits in (1) ?

- A. Sentence A
- B. Sentence B
- C. Sentence G
- D. Sentence H

Question: 19 of 30

Which one fits in (2) ?

- A. Sentence A
- B. Sentence B
- C. Sentence C
- D. Sentence D

Question: 20 of 30

Which one fits in (3) ?

- A. Sentence A
- B. Sentence D
- C. Sentence B
- D. Sentence E

Question: 21 of 30

Which one fits in (4) ?

- A. Sentence A
- B. Sentence B
- C. Sentence H
- D. Sentence E

Question: 22 of 30

Which one fits in (5) ?

- A. Sentence A
- B. Sentence B
- C. Sentence C
- D. Sentence D

Question: 23 of 30

Which one fits in (6) ?

- A. Sentence A
- B. Sentence F
- C. Sentence C
- D. Sentence D

Question: 24 of 30

Which one fits in (7) ?

- A. Sentence A
- B. Sentence G
- C. Sentence H
- D. Sentence C

Direction for Questions: 25 to 27

Read the poem below then answer the questions.

Sonnet
by **Elizabeth Bishop**

I am in need of music that would flow
Over my fretful, feeling finger-tips,
Over my bitter-tainted, trembling lips,
With melody, deep, clear, and liquid-slow.
Oh, for the healing swaying, old and low,
Of some song sung to rest the tired dead,
A song to fall like water on my head,
And over quivering limbs, dream flushed to glow!
There is a magic made by melody:
A spell of rest, and quiet breath, and cool
Heart, that sinks through fading colors deep
To the subaqueous stillness of the sea,
And floats forever in a moon-green pool,
Held in the arms of rhythm and of sleep.

Question: 25 of 30

In the poem, what does the speaker yearn for in the music?

- A. Loud and energetic melodies
- B. Slow and calming melodies
- C. Harsh and jarring melodies
- D. Chaotic and dissonant melodies

Question: 26 of 30

Which word in the poem emphasizes the sense of touch?

- A. Melody
- B. Fingertips
- C. Head
- D. Rhythm

Question: 27 of 30

What is the deeper meaning conveyed by the line, "A spell of rest, and quiet breath, and cool"?

- A. The speaker's desire for physical comfort
- B. The calming and healing power of music
- C. The presence of an actual spell in the poem
- D. The speaker's wish for a colder climate

Direction for Questions: 28 to 30

Read the four extracts below on the theme of human bonds.

Extract A:
Friendship is a precious bond that brings joy, support, and love into our lives. It is a relationship built on trust, understanding, and shared experiences. True friends are there for each other through thick and thin, offering a shoulder to lean on and a listening ear. They celebrate our successes and provide comfort during difficult times. Friendship is not limited by distance or time; it transcends boundaries and lasts a lifetime. A true friend is someone who accepts us for who we are, flaws and all, and encourages us to be the best version of ourselves. In a world full of uncertainties, friendship is a constant source of happiness and strength.

Extract B:
Helpful co-workers are the backbone of a successful and harmonious work environment. They are the ones who go above and beyond to lend a helping hand and make everyone's job easier. These individuals are selfless, reliable, and always willing to share their knowledge and expertise. They understand that collaboration and teamwork are key to achieving common goals. Helpful co-workers not only provide assistance with tasks, but they also offer guidance and support, boosting morale and fostering a positive work culture. Their presence creates a sense of unity and camaraderie, where everyone feels valued and appreciated. With helpful co-workers by your side, challenges become more manageable, and achievements become even more rewarding.

Extract C:
Having friends abroad is a truly enriching experience. It opens up a whole new world of perspectives, cultures, and traditions. These friendships transcend borders, allowing us to connect with people from different walks of life. Friends abroad offer a unique opportunity to learn about different customs, languages, and ways of living. They become our cultural ambassadors, helping us navigate unfamiliar territories and immersing us in their local experiences. These friendships also foster a sense of global citizenship, promoting empathy, understanding, and acceptance. Through our friends abroad, we gain a global network, creating lasting connections that may span continents but are bound by shared memories and a deep sense of friendship.

Extract D:
Companionship is a beautiful bond that connects individuals on a deeper level. It encompasses trust, understanding, and unwavering support. True companionship is not limited to romantic relationships; it extends to friends, family, and even pets. It is the joy of sharing laughter, the comfort in times of sorrow, and the strength in facing challenges together. Companionship brings a sense of belonging and fulfillment, as it nourishes our emotional well-being. It is the glue that holds communities together, fostering unity and empathy. In a world that often feels disconnected, companionship reminds us of the power of human connection and the beauty of shared experiences.

Question: 28 of 30

Which of the extracts discusses opening up a whole new world of perspectives, cultures, and traditions?

- A. Extract A
- B. Extract B
- C. Extract C
- D. Extract D

Question: 29 of 30

Which of the extracts refers to the glue that hold communities together?

- A. Extract A
- B. Extract B
- C. Extract C
- D. Extract D

Question: 30 of 30

Which of the extracts depicts the bond that makes the challenges more manageable?

- A. Extract A
- B. Extract B
- C. Extract C
- D. Extract D

Preptive Prepare Better
Series I Mathematical Reasoning Trial Test

Name: _____

Date: _____

Question: 1 of 35

Rishi drew lines to form triangles and stars.

Rishi formed a total of 10 triangles and stars. He drew 48 more lines for the stars than for the triangles. How many stars did he form?

○ **A.** 10 stars ○ **B.** 48 stars

○ **C.** 6 stars ○ **D.** 5 stars

Question: 2 of 35

There are some $2 and $5 notes in a box.
There are thrice as many $2 notes as $5 notes.
Given that the total amount of money in the box is $2255, how many $2 notes are there in the box?

○ **A.** 205 notes ○ **B.** 612 notes

○ **C.** 615 notes ○ **D.** 2255 notes

Question: 3 of 35

Yash bought sandwiches at the prices shown below.

Salmon Sandwiches

$ 9.60 each

Tuna Sandwiches

$ 6.40 each

Yash spent $464 on some salmon and tuna sandwiches. He bought 15 more salmon than tuna sandwiches. How many tuna sandwiches did Yash buy?

- A. 320
- B. 20
- C. 464
- D. 960

Question: 4 of 35

A box of pies costs $15 while a box of muffins costs $17. Each box contained either 4 pies or 3 muffins. Mrs Silva bought the same number of pies and muffins. She paid $207 more for the muffins.
How many boxes of pies did she buy?

- A. 27
- B. 9
- C. 23
- D. 45

Question: 5 of 35

The table shows the charges for bicycle rental.

Bicycle rental charges	
For the first hour	$8
For every additional $\frac{1}{2}$ hour or part thereof	$1.10

Shyam rented a bicycle from 8.30 a.m. to 11.45 a.m. How much did he pay for the rental of the bicycle?

- A. $13.50
- B. $5.50
- C. $58
- D. $3.15

Question: 6 of 35

Mrs Tan had two rolls of ribbons of the same length but different designs. She cut the first roll of ribbon into equal pieces of length 40 cm and there were 7 suns on each piece of ribbon as shown below.

First roll of ribbon

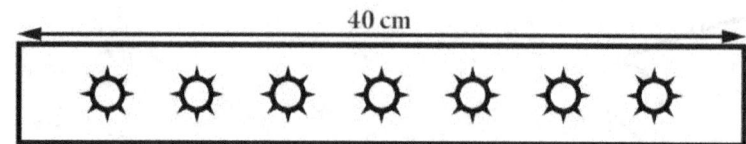

She then cut the second roll of ribbon into equal pieces of length 60 cm and there were 9 stars on each piece of ribbon as shown below.

Second roll of ribbon

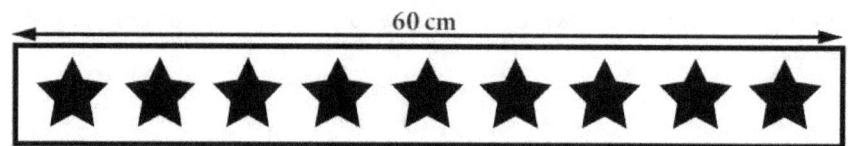

After she finished cutting both rolls of ribbons, she counted that the total number of suns was 126 more than the total number of stars. Find the length of one roll of ribbon.

- A. 555 cm
- B. 120 cm
- C. 3 cm
- D. 5040 cm

Question: 7 of 35

Henry gave Zoe $420.
He then spent $1/4$ of his remaining money on a bag.
As a result, he had $6/11$ of his money left.
How much money did Henry have at first?

- A. $1440
- B. $1450
- C. $1540
- D. $1550

Question: 8 of 35

Connie paid $27 for a belt after a discount at a bazaar. She then bought a skirt from another shop. She spent a total of $72 on these two items altogether.
She saved $12 total from discounts.
Given that Connie was given a 10% discount on the belt, find the percentage discount given for the skirt.
Round off your answer to one decimal place if necessary.

- A. 17.5%
- B. 17.6%
- C. 16.7%
- D. 16.5%

Question: 9 of 35

An apple juice factory's cylindrical tank is filled up before being transferred into a tanker. The graph below shows how the height of the apple juice in the tank changes.
How long did it take to empty the tank?

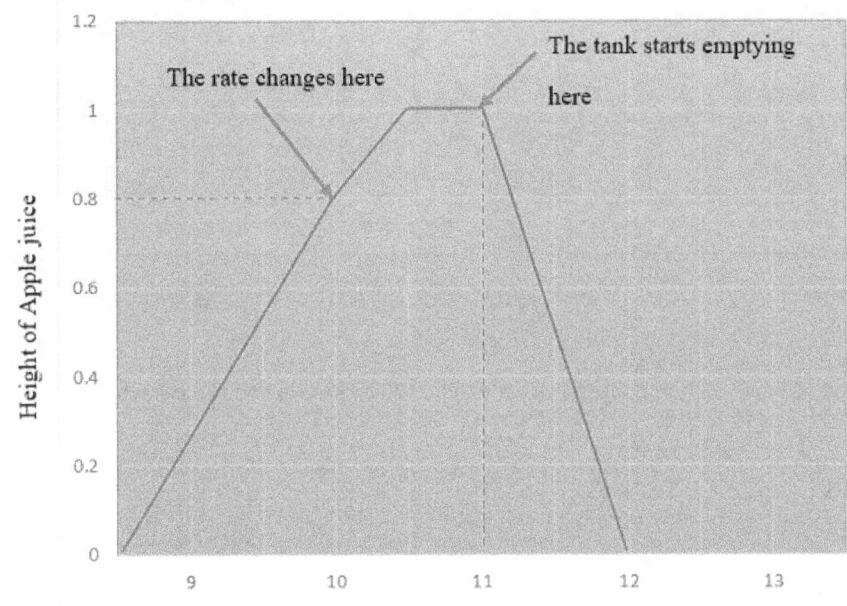

- A. 1.5 hours
- B. 2.5 hours
- C. 2 hours
- D. 1 hour

Question: 10 of 35

The stem and leaf diagram on the right shows the ages of some school teachers. How many teachers are in their forties?

```
3 | 3  5
4 | 0  5  7  8
5 | 1  4  9
6 | 1  3
Key: 5 | 4 = 54 years
```

- A. 2
- B. 4
- C. 6
- D. 3

Question: 11 of 35

A final mark for a grading examination is calculated from three components using the following formula:
Component A × 0.6 + Component B × 0.3 + Component C × 0.1.
What was this candidate's final approximate mark if the candidate obtained the following marks:

- Component A = 64
- Component B = 36
- Component C = 40

A. 76

B. 28

C. 64

D. 53

Question: 12 of 35

Four schools had the following proportion of pupils with special education needs. Which school had the lowest proportion of pupils with special education needs?

School	Proportion
P	2/9
Q	0.17
R	57 out of 300
S	18%

A. School Q

B. School R

C. School S

D. School P

Question: 13 of 35

Elena bought many types of toys at an average cost of $12. She then bought one of each of the following two toys and the average cost of all her toys became $14.

$28 $10

How many toys did she buy in total?

A. 7

B. 5

C. 6

D. 8

Question: 14 of 35

The diagram below shows a bar chart.

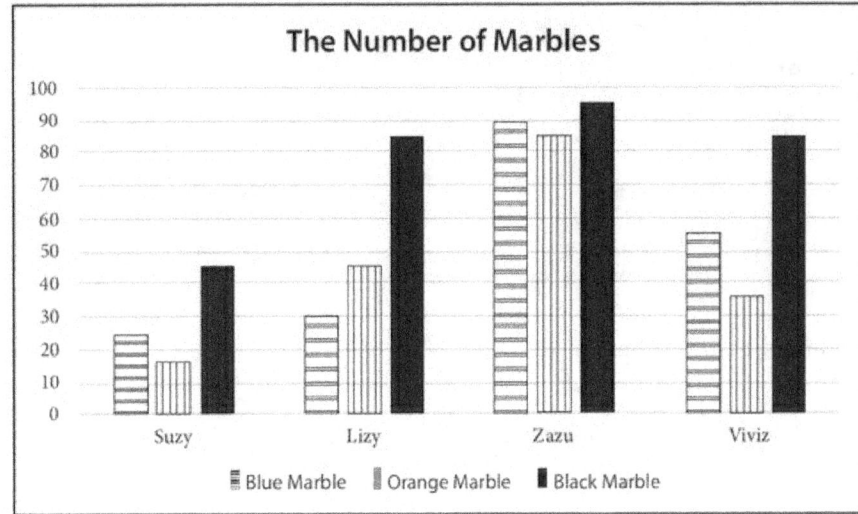

Which of the following is correct?

- A. The total number of blue marbles that Viviz and Lizy have is higher than the number of blue marbles that Zazu has.
- B. The total number of marbles that Suzy has is higher than the total number of Viviz's blue and grey marbles.
- C. Lizy has more marbles than Viviz.
- D. Zazu has the highest number of marbles.

Question: 15 of 35

In a trunk there are 5 chests; in each chest, there are 3 boxes; and in each box, there are 10 gold coins. The trunk, the chests, and the boxes are locked.

At least how many locks need to be opened to take out 50 coins?

- A. 6
- B. 8
- C. 7
- D. 9

Question: 16 of 35

The table shows the number of beads in a container.

Colour	Number of beads
White	150
Red	130
Blue	$\frac{2}{5}$ of the number of white beads

Daisy uses 3/5 white beads and 20% of red beads to embroider her dress.
Count the number of beads left in the container.

A. 225 B. 284
C. 224 D. 275

Question: 17 of 35

The diagram below shows 2 shapes.

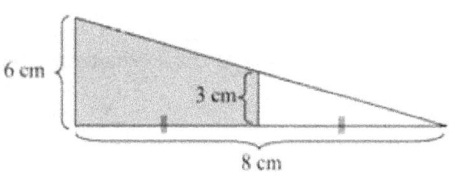 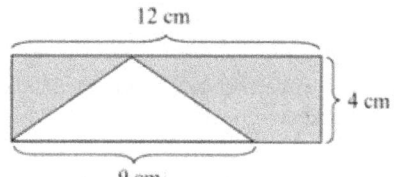

Find the total **non-shaded** area.

A. 35cm² B. 24cm²
C. 25cm² D. 15cm²

Question: 18 of 35

The diagram below shows measuring tools.

 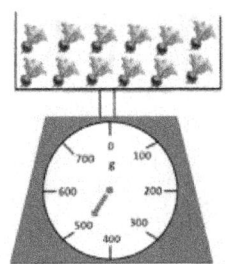

Rara buys 12 balls and 3 times beetroot shown in the picture above. How much is the total weight of all the items?

A. 1,055g
B. 1,536g
C. 1,533g
D. 1,305g

Question: 19 of 35

The diagram below shows measuring tools.

Fifi buys 8 apples. Find the weight of the apples.

A. 160g
B. 120g
C. 150g
D. 130g

Question: 20 of 35

The diagram below shows an equation.

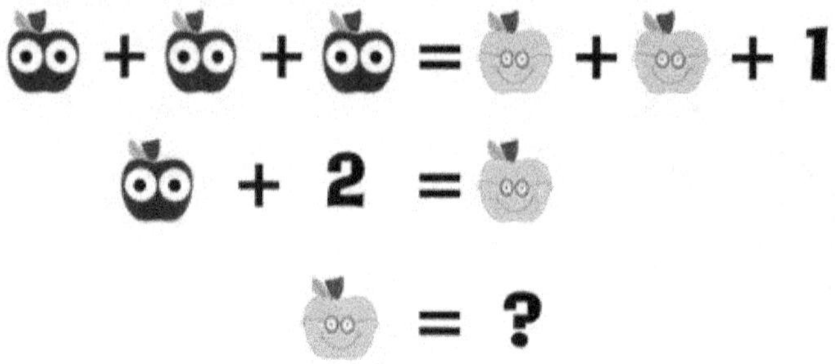

🍎 + 🍎 + 🍎 = 🍎 + 🍎 + 1

🍎 + 2 = 🍎

🍎 = ?

What is the answer?

- A. 5
- B. 7
- C. 6
- D. 8

Question: 21 of 35

204 eggs were packed into big and small cartons.
Each big carton contained 12 eggs and each small carton contained 6 eggs. There were 8 more big cartons than small cartons. How many such big cartons were used?

- A. 143
- B. 14
- C. 24
- D. 18

Question: 22 of 35

The following pictograph shows the number of books sold by a company during a week. Study the pictograph carefully and answer the question given below.

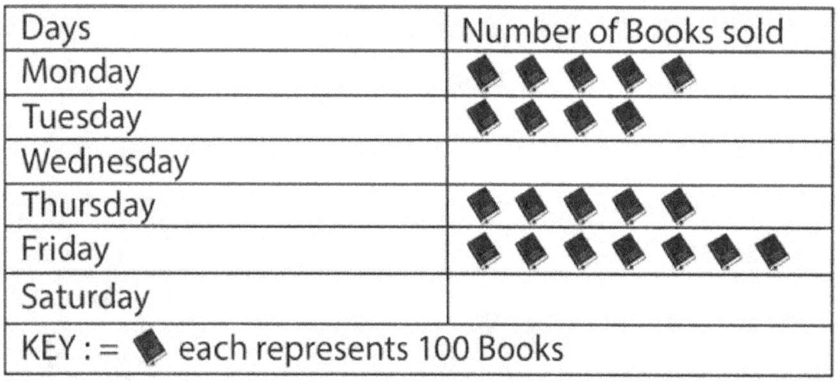

Days	Number of Books sold
Monday	📕 📕 📕 📕 📕
Tuesday	📕 📕 📕 📕
Wednesday	
Thursday	📕 📕 📕 📕
Friday	📕 📕 📕 📕 📕 📕
Saturday	
KEY : = 📕 each represents 100 Books	

Books sold on Wednesday are twice as many as those sold on Tuesday. Books sold on Saturday are 400 less than books sold on Wednesday. How many books were sold on **Tuesday and Saturday**?

- A. 600
- B. 800
- C. 260
- D. 300

Question: 23 of 35

Marlene was preparing for a race. She ran the same distance every day for five days. The graph shows the time taken, in minutes, to complete her run each day.

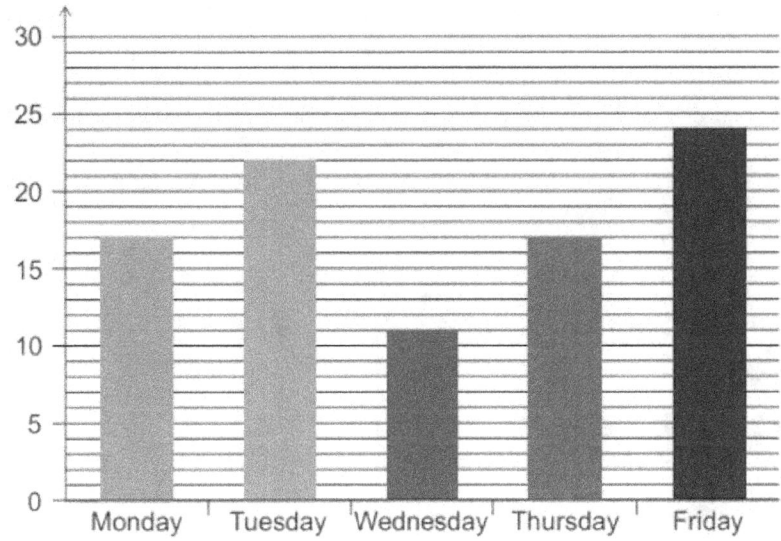

Which of these statement(s) is/are correct?

1. Marlene ran the fastest on Friday.
2. Marlene's time on Tuesday was exactly double of her time on Wednesday.
3. Marlene's total time running on Tuesday and Wednesday was the same as her total time running on Monday and Thursday.

- A. none of them
- B. Statement 1 only
- C. Statement 2 only
- D. Statement 3 only

Question: 24 of 35

The picture shows the exact number of shirts sold every day.

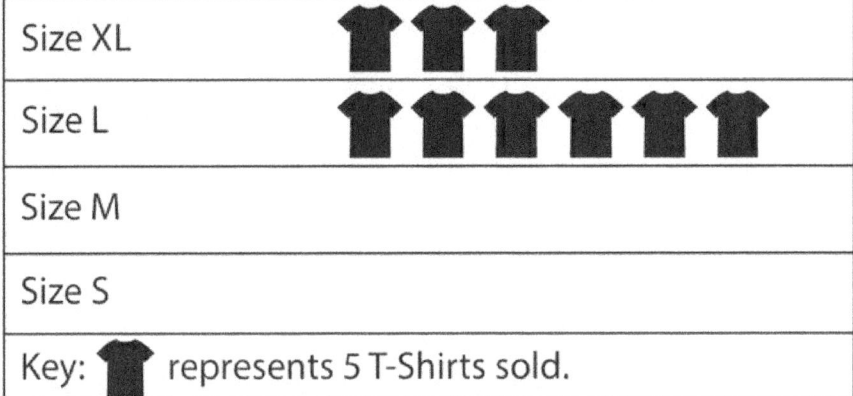

Shirts size S sold 2 times size XL, while shirt size M sold 3 times size S.
How many shirts were sold on **4 days**?

- A. 660
- B. 550
- C. 33
- D. 132

Question: 25 of 35

Lisa left her house at 8:30 AM and arrived at her friend's house at 10:15 AM. If the distance between their houses is 12 kilometers, what was Lisa's **approximate** average speed during her journey?

- A. 5 km/h
- B. 6 km/h
- C. 7 km/h
- D. 8 km/h

Question: 26 of 35

The pictograph shows the number of A.C. sets sold in several years.

Year	Number of A.C sets sold
1995	▮ ▮
2000	▮ ▮ ▮ ▮
2005	▮ ▮ ▮
2010	▮
2017	▮ ▮ ▮ ▮ ▮
KEY : Each ▮ represents 3000 A.C sets	

Which of the following is **wrong**?

- A. The difference between the number of A.C. sold in 2017 and 2000 is 2,500.
- B. The total number of A.C. sets sold in 2010 and 1995 is 9,000.
- C. The total of the number of A.C sets sold in 2017 and 2005 is 24,000
- D. None of the above.

Question: 27 of 35

Find the area of the following figure. Assume all triangles have the same dimensions.

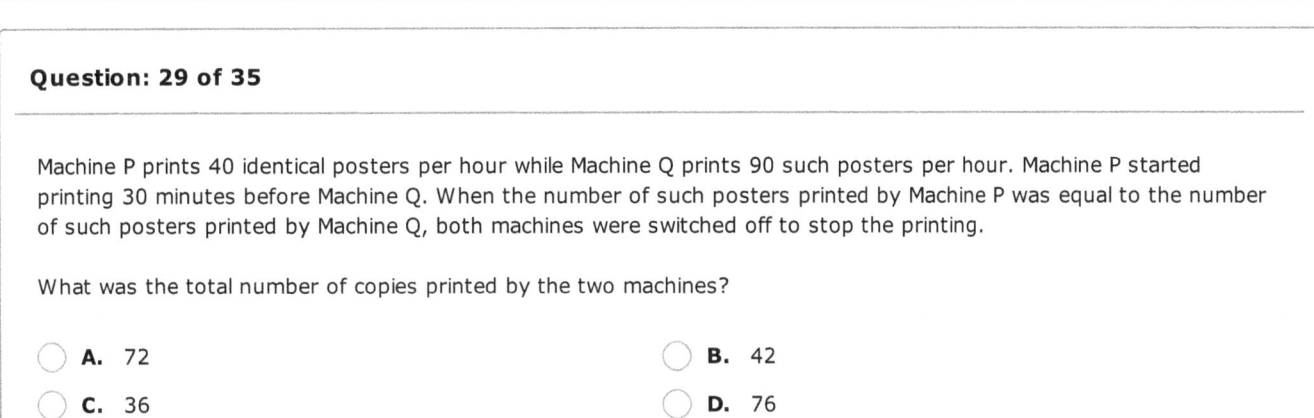

- A. 130 cm²
- B. 145 cm²
- C. 220 cm²
- D. 280 cm²

Question: 28 of 35

1 Tulip and a Rose cost $18
2 Roses and a Dahlia cost $25
3 Tulips and 2 Dahlia cost $34

What is the total value of 1 Tulip, 1 Dalia and 1 Rose?

- A. $20
- B. $23
- C. $25
- D. $27

Question: 29 of 35

Machine P prints 40 identical posters per hour while Machine Q prints 90 such posters per hour. Machine P started printing 30 minutes before Machine Q. When the number of such posters printed by Machine P was equal to the number of such posters printed by Machine Q, both machines were switched off to stop the printing.

What was the total number of copies printed by the two machines?

- A. 72
- B. 42
- C. 36
- D. 76

Question: 30 of 35

The average of four **3-digit numbers** is 500.
Two of the numbers are 150 and 230.
What is the **largest difference** between the other two numbers?

A. 348
B. 378
C. 358
D. 366

Question: 31 of 35

Connie started saving money in her piggy bank on a Friday. She saved $1 per day from Monday to Friday and $2 per day on Saturday and Sunday.
On which day of the week would she have saved $32 in her piggy bank?

A. Sunday
B. Tuesday
C. Friday
D. Saturday

Question: 32 of 35

Class X has 50% more students than Class Y. Class X has 40% fewer students than Class Z. Given that there are 80 students in the 3 classes, how many students are there in Class X?

A. 22
B. 24
C. 28
D. 32

Question: 33 of 35

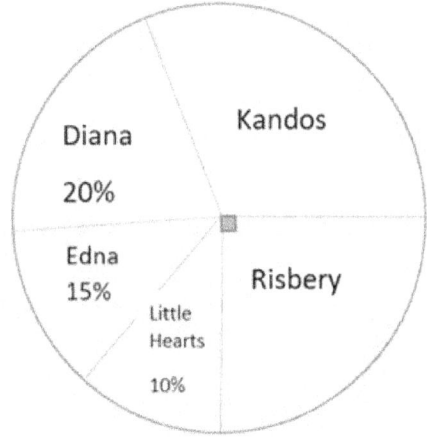

This chart shows favourite chocolate brands of 150 students.
Answer question using the above chart.

How many students like Kandos?

A. 30
B. 45
C. 55
D. Cannot be said

Question: 34 of 35

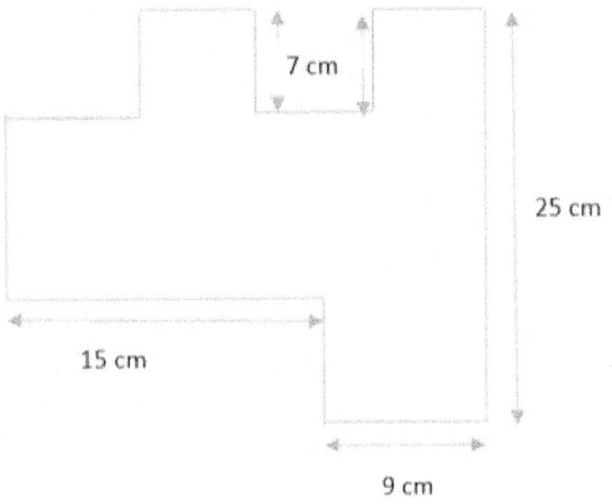

What is the perimeter of the given figure?

A. 98 cm
B. 100 cm
C. 112 cm
D. Data is not enough

Question: 35 of 35

The table shows the number of newspapers sold by Mr. Johan.

English Book	Number of sales
January	900
February	60 % more than the number of sales in January
March	20% lower than the number of sales in February
April	700

Count the number of newspapers sold by Mr. Johan in 4 months total.

A. 4,850
B. 4,459
C. 4,561
D. 4,192

Preptive Prepare Better
Series I Thinking Skill Trial Test

Name: _____

Date: _____

Question: 1 of 40

When Georgia was 10 years old, her parents promised to buy her a pet. That's because pets aren't allowed in the apartment complex they live in. Fortunately, they had some plans to relocate within a year. If she is 12 years old now, which of the following statements is possible?

   I. She has a pet.
   II. She doesn't have a pet.

- **A.** Statement I
- **B.** Statement II
- **C.** Both Statements
- **D.** None of the above

Question: 2 of 40

If you read one book by this author, there is a high chance of considering reading more of his work. That's according to research that shows 98% of people who have read one of his books have reacted in that manner. His way of captivating his readers is responsible for this behavior.

James: "Anyone who has read a book from this author has read more than one of them."
Graham: "One of the keys to get a huge fan base as an author is to captivate your readers."

Which of the following observations is correct?

- **A.** James Only
- **B.** Graham Only
- **C.** Both James and Graham
- **D.** Neither James nor Graham

Question: 3 of 40

Use the following statements to answer the question.
- Hunter will be visiting his grandmother after the school closes this weekend.
- He will only see his grandmother if his covid tests come back negative.
- Hunter hasn't seen his grandmother for years now.

After 2 weeks :

Rue: "The results were negative for Hunter since I saw a very recent photo of Hunter and his grandmother."
Jenna: "Schools have closed recently since Hunter's trip to her grandmother's house was scheduled one week back."

Which of the following reasoning is correct?

- **A.** Rue Only
- **B.** Jenna Only

Question: 4 of 40

A dice is numbered from 1 to 6 in different ways.
If 1 is adjacent to 6, 2, or 4, which of the following statements is definitely correct?

- A. 3 is opposite to 5
- B. 3 is adjacent to 5
- C. 1 is adjacent to 3
- D. 2 is opposite to 6

Question: 5 of 40

If you regularly hike in nature, going to the treadmill is unnecessary. Equally important, you don't need high-tech gadgets to exercise. The only thing necessary is a comfortable pair of hiking boots.

What is the best interpretation of the above statements?

- A. Hiking is better than using the treadmill for overall fitness.
- B. Using comfortable hiking boots is crucial for a successful workout.
- C. Hiking regularly is as effective as using the treadmill for fitness.
- D. Combining hiking and treadmill workouts is essential for a balanced exercise routine.

Question: 6 of 40

Different people had various opinions regarding the price of a laptop. A said that the price was between $800 and $900. B had a different range of between $750 and $850. C said that the price is less than or equal to $840.

If all the opinions were factual, which of the following conclusions is correct?

- A. The price can be $860
- B. The price can also be $920
- C. The price can be $820
- D. The price can be $880

Question: 7 of 40

In the music competition that started nine years ago, no participant has won the first place twice. Last season, Sophia claimed the victory, while Benjamin finished as the runner-up. According to this year's result, Sophia didn't break the tradition despite competing again. Benjamin also took another shot at the competition but did even worse than before. The only other contestants were Emma and Alexander.

If Alexander ended up in the last position, who won this time?

- A. Sophia
- B. Benjamin
- C. Emma
- D. Alexander

Question: 8 of 40

Use the following statements to answer the following question.

   I. Elsa is 49 years old.
   II. Nelson is her age mate.
   III. Elsa has brown hair.
   IV. Nelson has the same hair color.
   V. Elsa has long hair.

Which statements helps you know that Elsa and Nelson are identical twins?

- A. I, II, and III
- B. I, II, III, and IV
- C. All the above
- D. None of the above

Question: 9 of 40

A detective investigates a series of art thefts across Europe. The suspect leaves cryptic clues at each crime scene, each consisting of a geometric shape and a number. At the Louvre, the clue is a triangle and the number 10. At the Rijksmuseum, the clue is a pentagon and the number 15.

Based on this pattern, what shape and number should the detective expect at the next crime scene?

- A. Square, 20
- B. Heptagon, 20
- C. Circle, 30
- D. Impossible to determine

Question: 10 of 40

Roy loves music. Whenever he is in the mood for playing his favourite music, he plays multiple tracks. Today, he selected 3 of his tracks and put them in his 3 disc players to play them. The singles are of the same duration: 3 minutes. He started playing them at once. For how long did he enjoy the music play?

- A. 12 minutes
- B. 9 minutes
- C. 60 minutes
- D. 3 minutes

Question: 11 of 40

By their nature, their large number and the diversity of their operations/ sectors, family businesses are of great economic value in any economy. However, the sad news is that just a few percent of them barely survive past the third generation. Many family businesses get into problems for many reasons, including being badly affected by family conflicts. Also, weak or non-existent corporate governance, and favouritism in allocating resources, positions and appraisal are major challenges faced by family businesses.

If the given information is true, which of the following statements must be false?

- A. Family businesses are owned and run by family members.
- B. Many family businesses do not survive past the third generation.
- C. Family businesses do not have the challenges of weak corporate governance or authority allocation.
- D. Family conflicts most often adversely affect family businesses.

Question: 12 of 40

Three new pizzas, code-named A, B, and C are being developed by a restaurant.
A contains mushroom, spinach, cheese, tomatoes, red pepper and onion.
B contains cheese, spinach, tomatoes, red pepper and onion.
C contains mushrooms, cheese, tomatoes, onion and red pepper.

Which ingredients are included in only two?

- A. Mushroom and Spinach
- B. Cheese and Spinach
- C. Mushroom and Onions
- D. Tomatoes and Spinach

Question: 13 of 40

Consider the following notice at the supermarket during the Christmas week.

> **NOTICE.**
>
> The store will be opened this week on the following days and hours:
>
> - Weekdays: 9.00 am – 9.00 pm
> - Weekends: 8.00 am – 10.00 pm
> - Christmas Day: 1.00 pm to 7.00 pm
>
> Thanks for your understanding.
>
> **Management**

How many hours was the supermarket open that week, if the week starts on a Sunday and Christmas day was on Friday?

- A. 76
- B. 82
- C. 88
- D. 91

Question: 14 of 40

Trish is a famous painter.
Some of her exhibitions are packaged by her brother.

Which of the following is most likely correct, based on what we know so far?

- A. Trish has done some exhibitions as a famous painter
- B. Trish's brother is a businessman
- C. Trish's paintings are sold only at exhibitions
- D. Trish's brother loves promoting painting exhibitions

Question: 15 of 40

The Tennis competition started on the 13th April, Saturday.
Matches are only on weekend days and Wednesdays. There are 24 preliminary matches with 4 matches each match day. There are 12 Round-Two matches, with 3 matches each match day, the two semi-final matches were played on two match days. There was a rest day (in match day) before the final match. When was the final played?

- A. May 4th
- B. May 8th
- C. May 12th
- D. May 5th

Question: 16 of 40

The following tables show the time returned by five athletes in two races.
The athlete with the maximum improvement and completing Race 2 in equal to or less than 10.8 seconds will get an award.

Who got the award ?

Race 1

Colin	11.0 sec
David	11.2 sec
Gabriel	10.9 sec
Paul	11. 3 sec
Chris	11.1 sec

Race 2

Colin	11.0 sec
David	10.9 sec
Gabriel	10.8 sec
Paul	11.1 sec
Chris	10.8 sec

- A. Chris
- B. Paul
- C. Gabriel
- D. David

Question: 17 of 40

A library has three borrowing rules:

- You can only borrow maximum two books at a time.
- You cannot borrow fiction books and nonfiction books on the same day.
- You cannot borrow a book on the same day you return another book.

If you return a book on Monday, can you borrow a fiction book on Tuesday?

- A. Yes, always.
- B. Yes, if you do not borrow a non-fiction book on Tuesday
- C. No, never.
- D. It depends on whether you borrowed a fiction book on Friday.

Question: 18 of 40

You are trapped in a room with four doors labeled A, B, C, and D. Each door leads to a different outcome:

- Door A: You return to the start of the room.
- Door B: You teleport to a hidden chamber with another door leading back to the room.
- Door C: You escape the room and win, only if you come through another door.
- Door D: You encounter a riddle that, if solved correctly, opens Door C.

A sign tells you: "Only one door leads to freedom, others trap you further."

You can only choose one door. Which door should you choose?

- A. Door A
- B. Door B
- C. Door C
- D. Door D

Question: 19 of 40

You are driving on a road with three intersections, each controlled by a traffic light. Each light can be red, yellow, or green. You know that:

- At least one light is always green.
- No two consecutive lights are green.
- If one light is red, the next light must be yellow.

What is the possible combination of colors for the three lights, in order from left to right?

- A. Green, yellow, red
- B. Red, green, yellow
- C. Yellow, green, red
- D. Any combination is possible

Question: 20 of 40

A computer store assigns unique identification numbers to its laptops using the following system.

- The first digit represents the brand (1 for Dell, 2 for HP, 3 for Lenovo).
- The second digit represents the storage type (1 for SSD, 2 for HDD, others unknown).
- The remaining three digits represent the laptop's model number.

If a laptop's ID number is 23145, which of the following statements must be true?

- A. It is an HP laptop with an HDD.
- B. It is the 145th model of Lenovo laptops.
- C. It is a Dell laptop with an SSD.
- D. It is the 145th model of HP laptops.

Question: 21 of 40

Three friends, Alice, Bob, and Carol, provide information about their weekend plans:

Alice: "I am either going to the beach or the mountains."
Bob: "If Alice goes to the beach, I'll join her. Otherwise, I'll stay home."
Carol: "I'll join Bob wherever he goes."

If Carol stays home for the weekend, where is Alice most likely to be?

- A. Beach
- B. Mountains
- C. River
- D. Impossible to determine

Question: 22 of 40

You're hiking in a remote area and stumble upon a hidden cabin. Inside, you find a diary detailing a magician's life, ending with a cryptic entry: "**Farewell, stage. I vanish to where shadows dance.**"

Where might the magician have gone, based on the clue?

- A. A dark cave system shrouded in perpetual darkness.
- B. A secluded island surrounded by dense, sun-blocking foliage.
- C. A bustling city with a vibrant nightlife and hidden alleyways.
- D. An abandoned theater, their final performance space now consumed by darkness.

Question: 23 of 40

If the mirror is placed on the left line, then which of the given figures will be the correct image of the question figure.

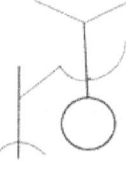

- A.
- B.
- C.
- D.

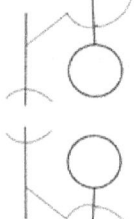

Question: 24 of 40

Gone are the days when Emma would comfortably go to work by bus. Since one of the buses broke down, people are scrambling for the remaining seats. Sometimes, the bus is full to capacity when it reaches the station where Emma takes her bus. Even when it isn't full, it is too crowded for Emma to enjoy her ride.

Based on the above information, what is likely true ?

- A. Emma should start taking the train instead.
- B. Emma will most likely look for another job to avoid this route.
- C. Since one of the buses broke down, Emma's commute has been inconvenienced.
- D. Many passengers are complaining about going by bus.

Question: 25 of 40

The code of a certain language is as follows :
126 means **Mary is good.**
425@ means **Good things are amazing.**
562 means **Bring good things**.

Which of the following is a possible code for *these things are good*?

- A. 257#
- B. 1234
- C. 54#$
- D. 425^

Question: 26 of 40

A store offers a special discount on specific items. The price tag shows the original price followed by a cryptic clue: "Subtract the product of the two digits". For example, a $40 item has the clue "75", implying $40 − (7 × 5) = $5.

You encounter three items with the following price tags and clues:

- Item A: $32, "24"
- Item B: $87, "51"
- Item C: $148, "69"

Which item has the highest discounted price after applying code ?

- A. Item A
- B. Item B
- C. Item C
- D. None of the item.

Question: 27 of 40

A renowned alchemist prepares a magical potion with four ingredients: Moonflower (M), Nightshade (N), Dragon Scale (D), and Phoenix Tear (P). Each ingredient possesses unique properties, and the order in which they're added determines the potion's final effect. Here's what we know:

- Adding M first leads to an invigorating potion.
- Adding D first leads to a tranquillizing potion.
- Adding N first leads to a truth-telling potion.
- Adding P first leads to a transformative potion.

Furthermore, certain ingredient combinations result in specific outcomes:

- M + N = poisonous
- (invisible) + M = hallucinogenic
- P + N = invisibility
- P + M = flight

Given the following sequence of ingredient additions: P -> N -> M, what will be the final effect of the potion?

- A. Invigorating and transformative
- B. Truth-telling and poisonous
- C. Hallucinogenic
- D. Impossible to determine with the given information

Question: 28 of 40

Three alien species use unique currencies: Zargs (Z), Glurps (G), and Sprockets (S). Their exchange rates are represented by inequalities:

Z > G
G > S

Based on these inequalities, we can determine ...

- A. 1 Z can be exchanged for 2 G.
- B. 1 G can be purchased with 3 S.
- C. You can always get more S by exchanging Z.
- D. More information is needed to determine the specific exchange rates.

Question: 29 of 40

You walk into a clothing store and find a beautiful jacket priced at $100. The store has a promotion: one item gets 50% off, another item gets 25% off, and any remaining items are full price. You also have a coupon for $10 off any purchase. How can you maximize your savings on the jacket?

- A. Buy the jacket only and use the coupon.
- B. Buy the jacket along with a cheap item and apply the 50% discount to the jacket.
- C. Buy the jacket along with another item and apply the 25% discount to the jacket.
- D. The savings are equal regardless of your purchase.

Question: 30 of 40

Tiffany is about going out on a winter day. She is perplexed about how to dress. She has the choice of wearing a blouse or shirt; a skirt or a pair of jeans or plain pants.
Because the weather is so cold she has the choice of wearing a pullover, a sweater, a parka or a blazer. She can then decide whether to wear a head warmer or a scarf.
In how many combinations can Tiffany dress from 'head to-toes' when going out?

- A. 36
- B. 48
- C. 24
- D. 12

Question: 31 of 40

Seats at a musical concert were arranged such that each successive row had 1 seat more than the one preceding it. In which row will 33 people be sitting if the first row has 16 people sitting?

- A. 19th
- B. 17th
- C. 15th
- D. 18th

Question: 32 of 40

In a scholarship test, Rob was placed ahead of Larry who scored 67%. Nim scored 60% but was not shortlisted for the interview. Out of four friends, Rob, Dan, Nim and Larry, only Rob and Dan, who were ahead of Nim were shortlisted for the interview.
With the information above, which one of the following is true?

- A. Nim was placed between Rob and Larry
- B. Dan scored more than 67%.
- C. Larry was placed between Rob and Dan
- D. Nim was placed third among the four friends

Question: 33 of 40

Audrey is faster than Hope, and Edith is faster than Audrey. This trio loves watching movies when together. It is Hope who often chooses the movie they will watch.
If all the above are facts, which of the following statements is also true?

- A. Hope is the fastest among the three individuals
- B. Audrey is the fastest among the three people
- C. Before watching a movie together, Hope often selects it
- D. Hope hates to watch a movie she hasn't chosen

Question: 34 of 40

If their memories serve them right, Ann, Gideon, and Angeline agree that their friend Eric resides in estate D. They also agree that his apartment is on the 4th floor. Two of them agree that the number after the hyphen is 3, whereas the one that follows is 5. Two of them also agree that the last number is 8.

If the apartment number comprises the estate name, floor number, a hyphen, and 4 numbers, which is the most probable number for Eric's apartment ?

- A. D4-3587
- B. D4-5218
- C. D4-4358
- D. D4-3568

Question: 35 of 40

Despite being a great event planner, Daniel has noticed that locals don't use his services often. It turns out that they think that Daniel won't accept gigs if the event is small.

If all the above are facts, which of the following statements can make him change this perspective?

- A. Relocating his offices and trying to woe a different clientele
- B. Using local channels such as the local magazine to advertise his business
- C. Advertising his business during huge global exhibitions and workshops
- D. Only offers discounts and coupons to his big clients

Question: 36 of 40

Find out that three figures from the given figures by which an equilateral triangle can be made.

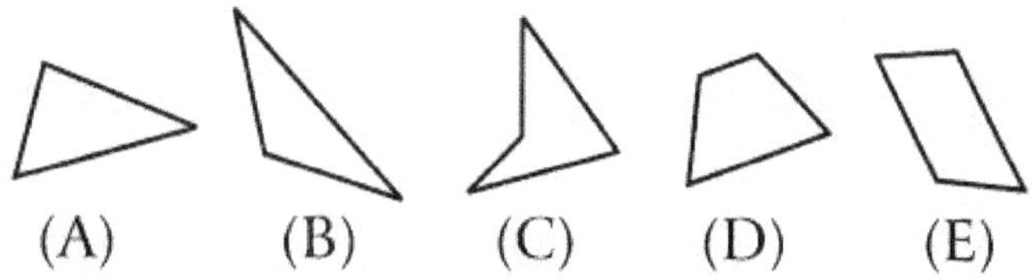

- A. BCD
- B. ACD
- C. CDE
- D. BDE

Question: 37 of 40

Replace question mark (?) with appropriate number from amongst the options.
48, 24, 42, 21, 36, 18, 30, ?, ?

- A. 15, 24
- B. 12, 21
- C. 18, 26
- D. 21, 25

Question: 38 of 40

Figure 1 shows a rectangular piece of paper.
The ratio of its length to its breadth is 4 : 3.
In Figure 2, the piece of paper is folded and cut along the dotted line.
Figure 3 shows the cut-out, C, and the remaining area of paper, R, which is square.

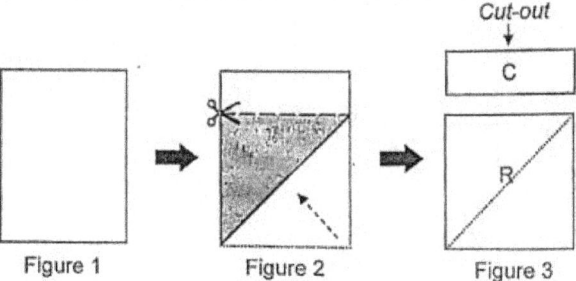

Figure 1   Figure 2   Figure 3

What is the ratio of the length to the breadth of C?

A. 4:9
B. 2:1
C. 3:1
D. 4:1

Question: 39 of 40

Use the rule of the pattern and find the missing number which will replace the question mark.

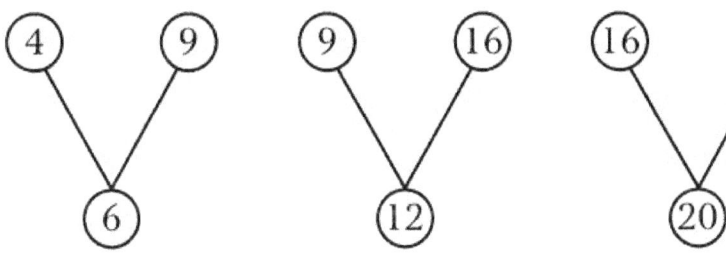

A. 21
B. 25
C. 45
D. 36

Find the missing figure from the given alternatives.

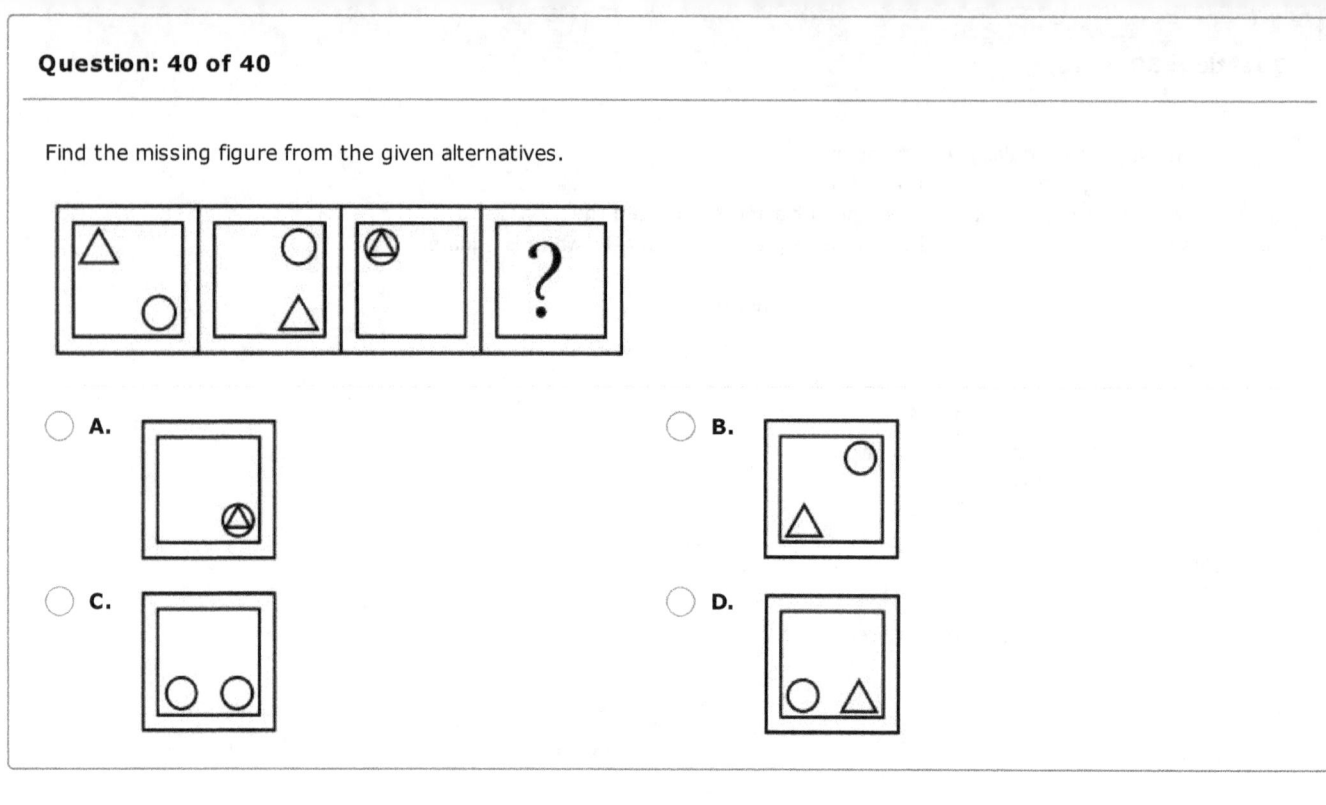

Preptive Prepare Better
Series J Reading Comprehension Trial Test

Name: _____

Date: _____

Direction for Questions: 1 to 5

Read the two extracts and then answer the questions.

Extract A: Fascinating Flamingoes: Nature's Pink Beauties

Flamingoes, with their striking pink plumage and unique characteristics, captivate the imagination of nature enthusiasts around the world. These graceful birds are renowned for their distinctive appearance and intriguing behaviors. In this extract, we delve into the fascinating world of flamingoes, exploring their habitat, physical features, and remarkable adaptations.

Flamingoes are predominantly found in tropical and subtropical regions, inhabiting shallow lakes, lagoons, and mudflats. These adaptable birds are known to thrive in a variety of habitats, ranging from coastal areas to inland wetlands. Their distribution spans across continents, with populations residing in Africa, Asia, Europe, and the Americas.

One of the most remarkable features of flamingoes is their vibrant pink plumage, which derives its color from the pigments in their food. Their long, slender legs are perfectly adapted for wading through shallow waters, while their distinctive curved beaks are specially designed for filter-feeding. Flamingoes also possess unique webbed feet, which enable them to paddle effortlessly through the water.

Flamingoes are highly social birds, often congregating in large flocks numbering in the thousands. These communal gatherings serve various purposes, including breeding, feeding, and protection against predators. Within these flocks, flamingoes exhibit complex social hierarchies, with dominant individuals often asserting their authority through displays of aggression or courtship rituals. One of the most fascinating aspects of flamingo behavior is their feeding technique. These birds are filter-feeders, using their specialized beaks to filter algae, small crustaceans, and other microscopic organisms from the water. By sweeping their beaks upside down through the water, flamingoes trap tiny food particles on tiny, hair-like structures called lamellae, located inside their beaks.

Flamingoes engage in elaborate courtship displays during the breeding season, with males and females performing synchronized dances and vocalizations to attract mates. Once paired, flamingoes construct intricate nests made of mud and vegetation, often in shallow waters to protect their eggs from predators. Both parents take turns incubating the eggs, which typically hatch after a month-long incubation period.

Despite their widespread distribution, flamingoes face numerous threats, including habitat loss, pollution, and disturbance from human activities. Conservation efforts are underway to protect flamingo populations and their habitats, including the establishment of protected areas and conservation initiatives aimed at mitigating human impacts.

Flamingoes are truly remarkable creatures, renowned for their distinctive appearance, social behavior, and unique adaptations. By studying and understanding these fascinating birds, we can gain valuable insights into the delicate balance of ecosystems and the importance of conservation efforts to preserve biodiversity for future generations.

Extract B: Conservation Challenges and Efforts

Flamingoes are masterful creatures when it comes to adaptation. Their unique physical characteristics enable them to thrive in diverse environments, from saltwater lagoons to alkaline lakes. One of the most striking features of flamingoes is their pink coloration, which comes from the pigments in the algae and crustaceans they consume. This specialized diet not only gives them their signature hue but also provides essential nutrients for their survival.

Beyond their color, flamingoes are renowned for their social behavior. These birds often gather in large flocks, numbering in the thousands, creating breathtaking displays as they move and feed together. Their synchronized movements and vocalizations serve important purposes, from signaling danger to attracting mates.

Flamingoes are also expert foragers, using their uniquely shaped bills to filter feed in shallow waters. Their curved bills are designed to trap tiny organisms like shrimp and plankton, allowing them to extract nutrients efficiently. This feeding behavior not only sustains the flamingo population but also plays a crucial role in maintaining the ecological balance of their habitats.

Despite their graceful appearance, flamingoes are formidable survivors. They have adapted to withstand extreme temperatures and harsh environmental conditions, demonstrating resilience in the face of adversity. From their specialized diet to their social structure and foraging techniques, flamingoes exemplify nature's ingenuity and beauty.

While flamingoes captivate with their beauty and grace, they also face numerous challenges in their natural habitats. Human activities such as pollution, habitat destruction, and climate change pose significant threats to flamingo populations worldwide.

One of the primary concerns for flamingoes is habitat loss. Wetlands, which serve as crucial breeding and feeding grounds for these birds, are disappearing at an alarming rate due to urbanization and agricultural expansion. Without adequate habitat, flamingoes struggle to find suitable nesting sites and food sources, leading to population declines.

Pollution is another major issue affecting flamingo populations. Chemical runoff from agricultural fields and industrial activities contaminates waterways, making them inhospitable for flamingoes and other wildlife. Plastic pollution is particularly harmful, as flamingoes often mistake plastic debris for food, leading to ingestion and digestive problems.

Climate change exacerbates these threats by altering the delicate balance of ecosystems. Rising temperatures and changes in precipitation patterns can disrupt the availability of food and water for flamingoes, forcing them to migrate in search of suitable habitats. Additionally, extreme weather events such as storms and droughts can devastate flamingo colonies, further endangering their survival.

Despite these challenges, conservation efforts are underway to protect flamingo populations and their habitats. Conservation organizations work to establish protected areas, regulate pollution, and promote sustainable land management practices. Public awareness campaigns raise awareness about the importance of preserving wetlands and reducing human impact on flamingo habitats.

Question: 1 of 30

How do flamingoes construct their nests, according to Extract A?

- A. Using twigs and branches
- B. Digging holes in the ground
- C. Making intricate nests of mud and vegetation
- D. Repurposing abandoned bird nests

Question: 2 of 30

How do conservation efforts aim to protect flamingo populations, according to Extract B?

- A. By promoting habitat destruction
- B. By increasing pollution levels
- C. Through establishing protected areas and regulating pollution
- D. By ignoring human impact on flamingo habitats

Question: 3 of 30

What demonstrates the resilience of flamingoes, as discussed in Extract B?

- A. Their ability to migrate long distances
- B. Their reliance on human intervention
- C. Their adaptation to extreme temperatures and harsh conditions
- D. Their dependency on artificial habitats

Question: 4 of 30

According to Extract A, what is the significance of the vibrant pink plumage of flamingoes?

- A. It serves as a defence mechanism against predators
- B. It provides camouflage in their natural habitats
- C. It derives its color from the pigments in their food and serves as a distinctive feature
- D. It indicates their age and maturity level

Question: 5 of 30

How do flamingoes utilize their uniquely shaped bills, as discussed in both extracts?

- A. For territorial displays
- B. For vocalizations and mating calls
- C. For foraging and filter-feeding
- D. For building nests and shelter

Direction for Questions: 6 to 10

Read the two extracts and then answer the questions.

Extract A: African Foods: A Culinary Journey Across the Continent

Africa, a continent teeming with diversity and vibrancy, boasts a rich culinary heritage that reflects its cultural tapestry, geographical landscapes, and historical influences. From the aromatic spices of North Africa to the hearty stews of West Africa and the exotic flavors of the East, African cuisine offers a tantalizing array of dishes that captivate the senses and ignite the palate. Embark on a culinary journey across the vast expanse of Africa and discover the flavors, traditions, and stories that define its diverse culinary landscape.

The culinary traditions of North Africa are shaped by the region's proximity to the Mediterranean Sea and its historical interactions with various cultures, including Arab, Berber, and Mediterranean influences. Staple ingredients such as couscous, lamb, olives, and aromatic spices like cumin, coriander, and cinnamon characterize North African cuisine. Signature dishes like tagine, a slow-cooked stew typically made with meat, vegetables, and dried fruits, showcase the region's emphasis on bold flavors and fragrant spices. Other notable dishes include falafel, hummus, and couscous served with tender lamb or succulent vegetables, reflecting the region's culinary diversity and culinary prowess.

West African cuisine is celebrated for its bold flavors, rich textures, and hearty dishes that reflect the region's agricultural abundance and cultural heritage. Staples such as rice, yams, plantains, and cassava form the foundation of West African cuisine, while ingredients like palm oil, peanuts, and spicy peppers infuse dishes with depth and complexity. Signature dishes such as jollof rice, a flavorful one-pot dish made with rice, tomatoes, and spices, and fufu, a starchy side dish typically served with soups or stews, exemplify the region's culinary ingenuity and sense of communal dining. Whether savoring a bowl of aromatic peanut soup or indulging in a plate of spicy grilled suya, West African cuisine offers a sensory feast that delights the palate and nourishes the soul.

The culinary landscape of East Africa is characterized by its diverse array of flavors, influenced by the region's cultural diversity and abundant natural resources. Staple ingredients such as grains, legumes, and fresh produce are complemented by aromatic spices like cardamom, cloves, and ginger, creating a symphony of flavors that dance across the palate. Signature dishes such as injera, a spongy flatbread served with savory stews and curries, and pilau, a fragrant rice dish flavored with spices and aromatics, reflect the region's culinary heritage and culinary creativity. Whether sampling the fiery flavors of Ethiopian cuisine or savoring the fragrant aromas of Swahili cuisine, East African cuisine offers a gastronomic adventure that celebrates the region's rich culinary heritage and cultural diversity.

Central African cuisine is characterized by its reliance on locally sourced ingredients, including root vegetables, leafy greens, and freshwater fish. Staples such as cassava, plantains, and maize are staples of the Central African diet, while ingredients like palm oil, peanuts, and chili peppers add depth and flavor to traditional dishes. Signature dishes such as moin moin, a steamed bean pudding, and fufu, a starchy side dish made from pounded cassava or plantains, reflect the region's emphasis on simple yet flavorful cooking techniques. Whether enjoying a bowl of hearty fish stew or savoring a plate of spicy grilled chicken, Central African cuisine offers a taste of the region's culinary heritage and culinary creativity.

In conclusion, African cuisine is a celebration of diversity, flavor, and tradition, offering a sensory journey that delights the palate and nourishes the soul. From the aromatic spices of North Africa to the bold flavors of West Africa and the exotic ingredients of East Africa, African cuisine reflects the continent's rich culinary heritage and cultural diversity. Whether savoring a traditional dish passed down through generations or exploring innovative fusion cuisine, African foods offer a culinary experience that is as diverse and vibrant as the continent itself.

Extract B: Why African Foods Stand Out: Celebrating Diversity, Flavor, and Tradition

African cuisine stands out on the global culinary stage for its rich tapestry of flavors, vibrant colors, and diverse culinary traditions that reflect the continent's cultural heritage and geographical richness. From the bustling markets of Lagos to the remote villages of the Sahara, African foods captivate the senses and ignite the palate with their unique blend of spices, textures, and aromas. Let's explore the reasons why African foods stand out and continue to captivate food enthusiasts around the world.

One of the most striking aspects of African cuisine is the incredible diversity of ingredients used in cooking. From hearty grains like millet and sorghum to exotic fruits like baobab and tamarind, Africa's bountiful landscapes offer a treasure trove of culinary delights. Each region boasts its own unique ingredients, spices, and cooking techniques, resulting in a rich tapestry of flavors and dishes that vary from one corner of the continent to another.

African cuisine is renowned for its bold and spicy flavors, which are achieved through the skillful use of aromatic spices, chili peppers, and herbs. Whether it's the fiery heat of West African jollof rice or the complex aromas of North African tagines, African dishes tantalize the taste buds with their vibrant and robust flavors. Spices like cumin, coriander, ginger, and paprika are often used to add depth and complexity to dishes, creating a sensory experience that is both exciting and satisfying.

African cooking places a strong emphasis on using fresh, seasonal ingredients that are locally sourced and harvested. From the colorful array of fruits and vegetables found in open-air markets to the freshly caught fish and seafood along the coastlines, African cuisine celebrates the bounty of nature and the importance of sustainable food practices. This focus on freshness not only enhances the flavor and nutritional value of dishes but also connects people to their local environments and traditions.

African cuisine is deeply rooted in cultural heritage and tradition, with recipes and cooking techniques passed down through generations. Many African dishes have symbolic significance and are prepared for special occasions, festivals, and celebrations. Whether it's the communal feasts of West Africa or the elaborate culinary rituals of Ethiopia's coffee ceremony, African food is imbued with meaning, history, and a sense of community that transcends borders and generations.

Despite its deep-rooted traditions, African cuisine is also characterized by its adaptability and innovation. Throughout history, African cooks have embraced new ingredients, cooking techniques, and cultural influences, resulting in a dynamic and ever-evolving culinary landscape. This spirit of creativity and experimentation is evident in dishes like South Africa's fusion cuisine, which blends traditional African flavors with European, Asian, and indigenous influences to create a unique and eclectic culinary experience.

African food is more than just sustenance; it is a celebration of community, hospitality, and togetherness. Meals are often shared with family and friends, with large gatherings and communal feasts serving as occasions for joy, bonding, and celebration. African hospitality is legendary, with guests welcomed with open arms and treated to generous portions of delicious food and warm hospitality.

In conclusion, African cuisine stands out on the global culinary stage for its diversity, flavor, and tradition. From the bold and spicy flavors of West Africa to the aromatic spices of North Africa and the fresh and seasonal ingredients of East Africa, African foods offer a sensory journey that celebrates the continent's rich cultural heritage and culinary creativity.

Question: 6 of 30

What culinary landscape is characterized by its reliance on locally sourced ingredients, as described in Extract A?

- A. North African cuisine
- B. West African cuisine
- C. East African cuisine
- D. Central African cuisine

Question: 7 of 30

According to Extract A, what is the overarching theme of African cuisine?

- A. Fusion of European and Asian influences
- B. Celebration of seasonal ingredients
- C. Emphasis on bold and spicy flavors
- D. Diversity, flavor, and tradition

Question: 8 of 30

According to Extract B, what contributes to the bold and spicy flavors of African dishes?

- A. Use of bland ingredients
- B. Incorporation of fresh herbs
- C. Skillful use of aromatic spices and chili peppers
- D. Absence of seasoning

Question: 9 of 30

What distinguishes the culinary traditions of North Africa, as described in Extract A?

- A. Emphasis on bold and spicy flavors
- B. Reliance on locally sourced ingredients
- C. Influence of Mediterranean and Arab cultures
- D. Preference for fusion cuisine

Question: 10 of 30

What role does communal dining play in the context of African cuisine, according to Extract B?

- A. It encourages individual dining experiences
- B. It limits the sharing of food among family and friends
- C. It fosters a sense of community, hospitality, and togetherness
- D. It discourages celebrations and festivals

Direction for Questions: 11 to 15

Read the text and then answer the questions.

Extract A:
Elephants, the majestic giants of the land, evoke an unparalleled sense of awe and admiration globally. Enigmatic and colossal, these creatures traverse the sprawling savannas and dense forests with an unparalleled grace, establishing profound familial bonds and showcasing remarkable cognitive abilities. Their social dynamics, reminiscent of our own societal structures, underscore the significance of community and nurturing the young among these gentle giants. Witnessing their sheer presence in the wild sparks a profound wonder, a poignant reminder of the intricate, mesmerizing tapestry of life within the animal kingdom.
Their colossal forms, meandering through landscapes with a gentle grandeur, serve as a testament to the beauty and complexity inherent in the natural world. Elephants stand as poignant reminders of resilience, unity, and the harmonious coexistence that underscores their existence in the untamed wild.

Extract B:
Dolphins, the captivating ambassadors of the sea, embody an exquisite blend of grace and intellectual prowess that enchants all who encounter them. Their playful demeanor and extraordinary agility paint a mesmerizing portrait of life beneath the ocean's shimmering surface. Often found traversing the seas in harmonious pods, dolphins' social intricacies, communicated through a symphony of sounds, reflect a level of intelligence that resonates with our own societal complexities.
Their acrobatic leaps and synchronized movements transcend mere spectacle, offering a poignant glimpse into the harmony that defines their underwater world. Dolphins symbolize the vibrant vitality thriving within the ocean's depths, underscoring the urgent need to safeguard their habitats and protect the intricate ecosystems they inhabit.

Extract C:
Owls, enigmatic creatures of the night, embody an aura of mystique and wisdom that transcends the realms of the ordinary. Navigating the veil of darkness with unparalleled finesse, these nocturnal beings exude an air of ancient knowledge, finely honed instincts, and a remarkable ability to thrive under the moon's luminous gaze. Revered across diverse cultures as symbols of foresight and intuition, owls command admiration for their unparalleled sight that reveals what remains hidden to others.
Their silent, almost ethereal flight and penetrating gaze evoke a sense of reverence, inviting humanity to embrace the enigmatic facets of the natural world and foster a deep respect for the delicate balance within the ecosystems they inhabit.

Extract D:
Butterflies, nature's ephemeral marvels, encapsulate the profound concepts of transformation and sheer beauty. These delicate beings undergo a breathtaking metamorphosis, a miraculous journey from unassuming caterpillars to resplendent creatures adorned with vibrant hues and intricate patterns. Fluttering gracefully through verdant gardens and sun-kissed meadows, butterflies scatter joy with their ephemeral flights, leaving fleeting traces of their exquisite existence.
Their transient lifespan serves as a poignant allegory for the transient nature of beauty itself, urging us to embrace the present moment and cherish the ephemeral splendor that graces our lives. Within nature's intricate tapestry, butterflies symbolize the cyclical rhythms of life, serving as timeless reminders to embrace change and revel in the fleeting moments of sublime magnificence.

Question: 11 of 30

In which extract do animals navigate the seas in harmonious pods, showcasing social intricacies and a level of intelligence akin to human societal complexities?

- A. Extract A
- B. Extract B
- C. Extract C
- D. Extract D

Question: 12 of 30

Which extract features animals revered across diverse cultures as symbols of foresight and intuition, with an ability to reveal what remains hidden to others?

- A. Extract A
- B. Extract B
- C. Extract C
- D. Extract D

Question: 13 of 30

In which extract do animals undergo a breathtaking metamorphosis, transforming from unassuming caterpillars to resplendent creatures with vibrant hues and intricate patterns?

- A. Extract A
- B. Extract B
- C. Extract C
- D. Extract D

Question: 14 of 30

Which extract describes animals that flutter gracefully through gardens and meadows, scattering joy with their ephemeral flights and leaving traces of exquisite existence?

- A. Extract A
- B. Extract B
- C. Extract C
- D. Extract D

Question: 15 of 30

In which extract do animals symbolize the cyclical rhythms of life and serve as timeless reminders to embrace change and revel in fleeting moments of sublime magnificence?

- A. Extract A
- B. Extract B
- C. Extract C
- D. Extract D

Direction for Questions: 16 to 20

Read the poem and then answer the questions.

Be Thou My Vision
by Dallan Forgaill

Be Thou my vision, O Lord of my heart;
Naught be all else to me, save that Thou art;
Thou my best thought, by day or by night,
Waking or sleeping, Thy presence my light.
Be Thou my wisdom, and Thou my true word;
I ever with Thee and Thou with me, Lord;
Thou my great Father, I Thy true son;
Thou in me dwelling, and I with Thee one.
Be Thou my battle shield, sword for the fight;
Be Thou my dignity, Thou my delight;
Thou my soul's shelter, Thou my high tower:
Raise Thou me heav'nward, O power of my power.
Riches I heed not, nor man's empty praise,
Thou mine inheritance, now and always:
Thou and Thou only, first in my heart,
High King of Heaven, my treasure Thou art.
High King of Heaven, my victory won,
May I reach heaven's joys, O bright heaven's Sun!
Heart of my own heart, whatever befall,

Still be my vision, O Ruler of all.

Question: 16 of 30

What does the phrase "Be Thou my vision" signify in the poem?

- A. Seeking clarity of sight
- B. Yearning for material success
- C. A wish for physical strength
- D. Longing for artistic inspiration

Question: 17 of 30

What does the speaker imply by calling the Lord "Thou my great Father, I Thy true son"?

- A. The speaker seeks protection from their parents
- B. The speaker desires material inheritance
- C. The speaker acknowledges a spiritual relationship
- D. The speaker wishes for earthly comforts

Question: 18 of 30

What sentiment does the line "Thou my soul's shelter, Thou my high tower" convey?

- A. Seeking refuge and protection in the divine
- B. A desire for physical strength
- C. The speaker's admiration for nature
- D. Longing for financial stability

Question: 19 of 30

How does the speaker view worldly possessions and praise from others?

- A. They are considered valuable and sought after
- B. They are disregarded as unimportant
- C. They are seen as essential for happiness
- D. They are sought after but with caution

Question: 20 of 30

What emotion does the concluding line "Still be my vision, O Ruler of all" evoke?

- A. Fear
- B. Serenity
- C. Sadness
- D. Confusion

Direction for Questions: 21 to 27

Seven sentences have been removed from the text. Choose from the sentences (A – H) the one which fits each gap (1 – 7). There is one extra sentence which you do not need to use.

The Mysterious Harappa Civilization

The Harappa civilization, also known as the Indus Valley civilization, flourished around 2600 to 1900 BCE in what is now present-day Pakistan and northwest India. **1...** The Harappan people displayed advanced urban planning, remarkable engineering skills, and a sophisticated system of writing that is yet to be fully deciphered. **2...** One of the most remarkable features of these cities was their advanced sewage and drainage systems. The well-organized streets and houses were built using standardized bricks, suggesting a centralized authority. The city of Harappa, for which the civilization is named, had fortified walls and a citadel, indicating the presence of a ruling elite. Archaeological excavations at Harappa and other Harappan sites have revealed a wealth of artifacts, including pottery, jewelry, and seals. **3...** The Harappans were skilled in metallurgy, as evidenced by their copper and bronze tools and ornaments. **4....** Despite numerous attempts, scholars have not been able to fully decipher this script, making it one of the greatest unsolved mysteries of ancient history. The Indus script was found on seals, pottery, and other artifacts, suggesting a widespread use of writing for administrative and commercial purposes. **5...** Some scholars speculate that environmental factors, such as a shift in the course of the Indus River or a change in climate, may have contributed to its downfall. Others propose the possibility of invasion or internal conflicts. Whatever the cause, the Harappa civilization gradually declined, and by around 1900 BCE, it had ceased to exist. In conclusion, the Harappa civilization was a highly advanced and prosperous civilization that left behind a rich archaeological legacy. **6...** The Harappa civilization serves as a reminder of the ancient civilizations that thrived in the Indus Valley thousands of years ago. **7...**

A. The Harappa civilization was characterized by well-planned cities, with their layout based on a grid pattern.
B. These seals, made of steatite, depict various animals and symbols and are believed to have been used for trade and administrative purposes.
C. The decline of the Harappa civilization remains a subject of debate among historians
D. The well-planned cities, advanced engineering, and enigmatic writing system of the Harappans continue to captivate the imagination of historians and archaeologists.
E. One of the most intriguing aspects of the Harappa civilization is its system of writing, known as the Indus script.
F. This ancient civilization remains one of the most intriguing and enigmatic. archaeological discoveries of our time.
G. The city of Harappa, for which the civilization is named, had fortified walls and a citadel, indicating the presence of a ruling elite.
H. It leaves behind a lasting impact on human history.

Question: 21 of 30

Which one fits in (1) ?

- A. Sentence A
- B. Sentence B
- C. Sentence G
- D. Sentence H

Question: 22 of 30

Which one fits in (2) ?

- A. Sentence A
- B. Sentence B
- C. Sentence C
- D. Sentence D

Question: 23 of 30

Which one fits in (3) ?

- A. Sentence A
- B. Sentence D
- C. Sentence B
- D. Sentence E

Question: 24 of 30

Which one fits in (4) ?

- A. Sentence A
- B. Sentence B
- C. Sentence H
- D. Sentence E

Question: 25 of 30

Which one fits in (5) ?

- A. Sentence A
- B. Sentence B
- C. Sentence C
- D. Sentence D

Question: 26 of 30

Which one fits in (6) ?

- A. Sentence A
- B. Sentence B
- C. Sentence C
- D. Sentence D

Question: 27 of 30

Which one fits in (7) ?

- A. Sentence A
- B. Sentence G
- C. Sentence H
- D. Sentence C

Direction for Questions: 28 to 30

Read the poem and choose the correct answers.

O Captain! My Captain!
by **Walt Whitman**

O Captain! My Captain! our fearful trip is done;
The ship has weather'd every rack, the prize we sought is won;
The port is near, the bells I hear, the people all exulting,
While follow eyes the steady keel, the vessel grim and daring:
But O heart! heart! heart!
O the bleeding drops of red,
Where on the deck my Captain lies,
Fallen cold and dead.
O Captain! my Captain! rise up and hear the bells;
Rise up—for you the flag is flung—for you the bugle trills;
For you bouquets and ribbon'd wreaths—for you the shores a-crowding;
For you they call, the swaying mass, their eager faces turning;
O captain! dear father!
This arm beneath your head;
It is some dream that on the deck,
You've fallen cold and dead.

Question: 28 of 30

What is the significance of the line "the prize we sought is won"?

- A. Goal of the journey has been achieved
- B. The journey has ended
- C. The journey is tiresome
- D. None of the above

Question: 29 of 30

Why does the speaker refer to the ship as "the vessel grim and daring"?

- A. Ship's resilience and courage
- B. The weaknesses of the ship
- C. The long journey of the ship
- D. None of the above

Question: 30 of 30

Why does the speaker call the captain "dear father"?

- A. They are strangers
- B. To convey an emotional connection
- C. They have known each other
- D. None of the above

Preptive Prepare Better
Series J Mathematical Reasoning Trial Test

Name: _____

Date: _____

Question: 1 of 35

The total cost of 5 shirts and 6 pants is $210.55.
The total cost of 3 shirts and 2 pants is $91.45.
All the shirts are identical and all the pants are identical.

Find the cost of 1 shirt.

- A. $15.95
- B. $63.80
- C. $15.90
- D. $274.35

Question: 2 of 35

Sam had some apples. He sold 210 apples in the afternoon and $\frac{3}{7}$ of the remaining apples in the evening.

In the end, he had $\frac{1}{3}$ of the total number of apples left.

How many apples did he have at first?

- A. 505 apples
- B. 504 apples
- C. 210 apples
- D. 42 apples

Question: 3 of 35

The table below shows the number of girls and boys who wear and do not wear spectacles in Class 6W.

	Number of girls	Number of boys
Wear spectacles	8	?
Do not wear spectacles	7	11

The **ratio** of the number of girls to the number of boys in 6W is 3 : 5.

Find the number of boys who wear spectacles in 6W.

- A. 5 boys
- B. 25 boys
- C. 14 boys
- D. 15 boys

Question: 4 of 35

The price of 1 Dino melon was $20. A shopkeeper gave 1 Dino melon free for every 4 Dino melon that were bought. Lisa spent a total of $620 on the Dino melons.
How many Dino melons did Lisa get?

A. 38
B. 31
C. 7
D. 20

Question: 5 of 35

At a carnival, the children were put into two groups. $\frac{2}{3}$ of Group A were boys and $\frac{3}{5}$ of Group B were girls. Group B had twice as many children as Group A.
There were 26 fewer boys in Group A than Group B.

What fraction of the total number of children at the carnival were boys? Give your answer in the simplest form.

A. 2/5
B. 5/6
C. 22/45
D. 8/45

Question: 6 of 35

Janet spent 30% of her money on 7 cupcakes and 4 cookies on Monday.
The cost of each cupcake was twice the cost of each cookie.
She bought some more cupcakes and another 10 cookies with 6/7 of her remaining money on Tuesday.
How many cupcakes did she buy on Tuesday?

A. 26
B. 13
C. 25
D. 28

Question: 7 of 35

Adrian and Bert had a total of $4563. After Adrian spent $\frac{1}{4}$ of his money and Bert spent $\frac{2}{3}$ of his money, they had an equal amount of money left.
How much money was Bert left with?

A. $1053
B. $1035
C. $1005
D. $1030

Question: 8 of 35

The picture shows the duration of rain in a region.

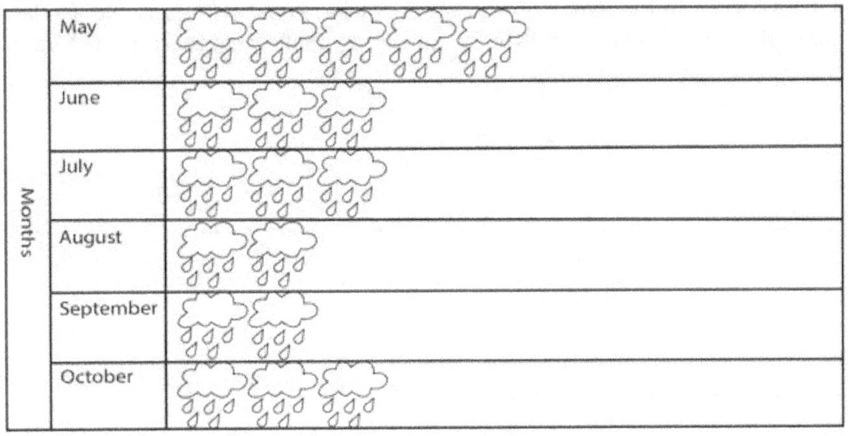

Months	May	☁☁☁☁☁
	June	☁☁☁
	July	☁☁☁☁
	August	☁☁
	September	☁☁
	October	☁☁☁

☁ = 50 minutes

Which of the following is a correct statement?

- A. There are 100 more minutes of rain in August than in September.
- B. There are 150 more minutes of rain in October than in May.
- C. There are 100 fewer minutes of rain in May than in June
- D. There are 50 more minutes of rain in June than in August.

Question: 9 of 35

Jack has $y for pocket money. Krishan has thrice as much pocket money as Jack. Latiff has $10 less than Krishan.

What is the total amount of pocket money the three boys have in terms of y?

- A. $(7y-10)
- B. $(7y-20)
- C. $(8y-6)
- D. $ (9y-9)

Question: 10 of 35

Adam has three fewer one-dollar coins than ten-cent coins. The total value of the ten-cent coins is $7.80 less than the total value of the one-dollar coins.
How many coins does Adam have in all?

- A. 22
- B. 20
- C. 21
- D. 24

Question: 11 of 35

In a library $\frac{3}{5}$ of the books were English books and $\frac{3}{8}$ of the remaining books were Chinese books. The rest were Malay books. There were 133 more English books than Malay books in the library. How many books were there in the library altogether?

- A. 380
- B. 228
- C. 308
- D. 448

Question: 12 of 35

The angles of a quadrilateral are $(p+25)°$, $2p°$, $(2p-15)°$ and $(p+20)°$.
What is the value of the largest angle?

- A. 105°
- B. 110°
- C. 115°
- D. 135°

Question: 13 of 35

What are the correct coordinates of the vertices of the given polygon?

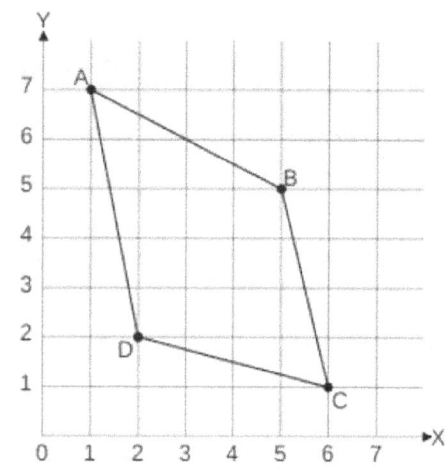

- A. A(1, 4) B(5, 5) C(6, 1) D(2, 2)
- B. A(1, 7) B(5, 5) C(6, 1) D(3, 2)
- C. A(1, 7) B(5, 5) C(6, 1) D(2, 2)
- D. A(1, 7) B(5, 4) C(6, 1) D(2, 2)

Question: 14 of 35

Which of the following statements is correct?

- A. $(2^3)^2$ and $(3^2)4^4$ are not the same
- B. $(2^3)^2$ and $(3^4)^2$ are the same
- C. $(6^{49})^2 = 6^{492}$
- D. $(8^7)^3 = 8^{10}$

Question: 15 of 35

The librarian spent $1888 buying books for the school library during the promotion as shown below. How many books did she buy altogether?

Books
At $16 each

PROMOTION
Buy any 18 books
and
get the 19th and 20th book at
half price each

- A. 124
- B. 146
- C. 144
- D. 68

Question: 16 of 35

Consider the following 3 statements.

1. Price of 4 toffees is $2
2. Price of 1 chocolate and 2 lollipops is $3.4
3. When we subtract price of two lollipops from the price of a toffee, the remaining price is equal to ½ price of a lollipop.

Calculate the total amount you have to pay if you buy 1 toffee, 1 chocolate and a lollipop.

- A. $2.10
- B. $3.70
- C. $4.50
- D. $5.20

Question: 17 of 35

AC and DF are parallel lines. What is the value of ABX angle?

- A. 65°
- B. 115°
- C. 90°
- D. 105°

Question: 18 of 35

In a farm there are 68 rabbits, 95 cattle and the number of hens is 25% of number of rabbits. If $1/34$ rabbits and 20% percent of cattle are transferred to another B farm, how many animals remain in Farm A?

- A. 160
- B. 159
- C. 158
- D. 157

Question: 19 of 35

The picture shows an automatic fish feeder and its specification.

Automatic fish feeder

Specifications
- Container size: Moderate
- Dried Food and pellet maybe used
- A timer is used to arrange feeding time
- Use the latest technology to prevent food from getting moist or stuck in the container
- Can be operated manually or automatically
- Digital screen display

If Eng Wei decides to feed the fish 4 times a day in equal interval with the first feeding time at 7:35 a.m., at what time should he feed the fish for the third feeding?

- A. Fishes are fed for the third time at 7:45 p.m.
- B. Fishes are fed for the third time at 7:35 p.m.
- C. Fishes are fed for the third time at 7:25 p.m.
- D. Fishes are fed for the third time at 7:15 p.m.

Question: 20 of 35

In the diagram below, M' is the image of M in an axis of reflection.

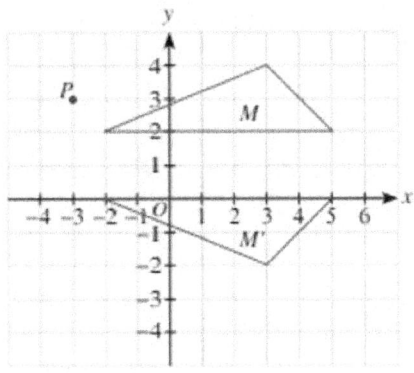

Determine the coordinates of P' under the same axis of reflection.

A. (-3 , -1)
B. (3 , 1)
C. (-3 , 1)
D. (3 , -1)

Question: 21 of 35

Mrs Lin prepared 160 chicken wings and some nuggets for a party. At one point during the party, an equal number of chicken wings and nuggets were eaten. 25% of the chicken wings and 20% of the nuggets were left. She then increased the number of chicken wings. After that, there was a total of 65 chicken wings.

How many nuggets did Mrs Lin prepare for the party?

A. 15
B. 150
C. 120
D. 130

Question: 22 of 35

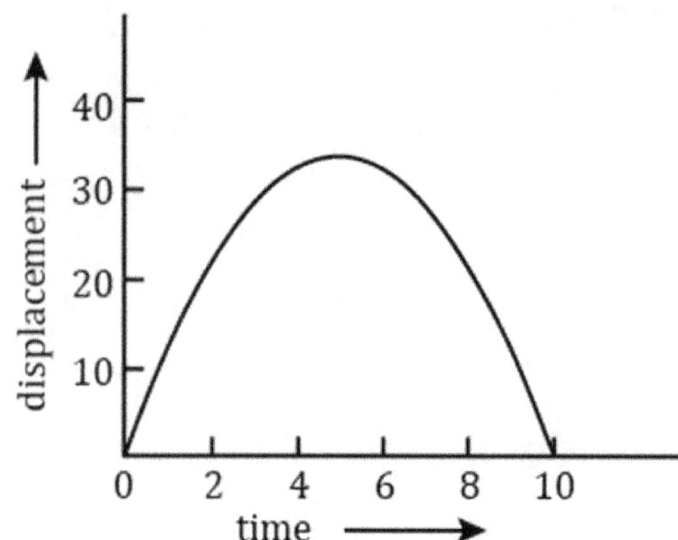

Layla throws a ball. The above graph represents the displacement- time graph of the ball (metre and sec). When it has a displacement of 20 m what is/are the times relevant for that displacement?

- A. 2s
- B. 4s
- C. 2s and 8s
- D. 8s

Question: 23 of 35

Which of the following statements is/ are **incorrect**?

X. $4/5 + 1/5 + 2/5$ is more than 1
Y. $1/8 - 1/7$ gives a negative number
Z. $8/9$ is less than $6/7$

- A. X and Y only
- B. Y and Z only
- C. Y only
- D. Z only

Question: 24 of 35

Jessie had $80 less than Vincent at first. Vincent gave Jessie $60. The ratio of Jessie's money to Vincent's money now became 11 : 7.
How much did Vincent have at first?

- A. $100
- B. $110
- C. $130
- D. $150

Question: 25 of 35

The figure on the right shows a 'star'. Find ∠x.

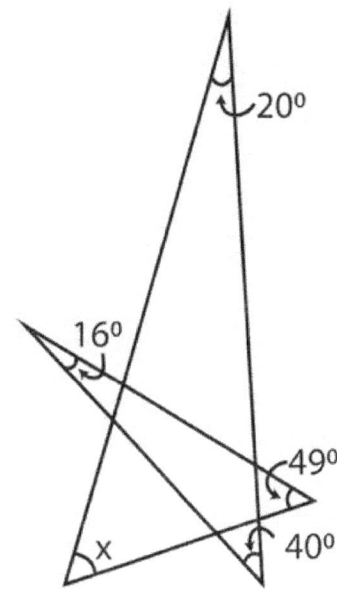

A. 65°
B. 55°
C. 45°
D. 75°

Question: 26 of 35

Jason had 40% fewer coins than Tom. Jason gave his sister some of his coins. As a result, the number of coins he had decreased by 25%.
How many coins did Jason have left if both had 58 coins altogether at the end?

A. 18
B. 16
C. 15
D. 12

Question: 27 of 35

What is the perimeter of the given figure?

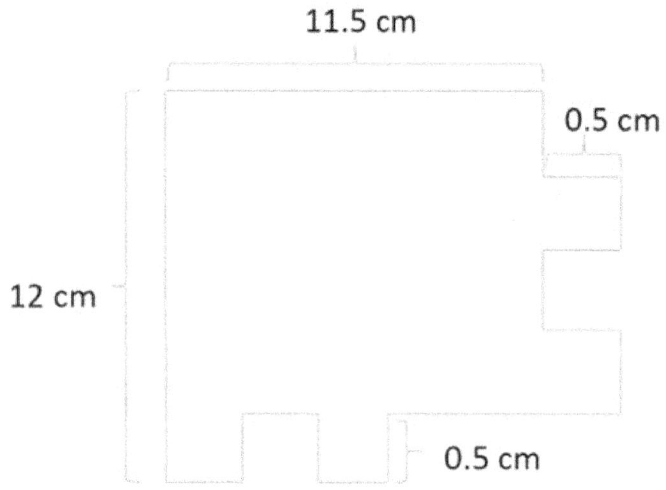

[Figure with measurements: 11.5 cm (top), 0.5 cm (right notch), 12 cm (left side), 0.5 cm (bottom notch)]

- A. 54 cm
- B. 52.5 cm
- C. 50 cm
- D. 48 cm

Question: 28 of 35

An ice cream cone costs $1.60 each. Each customer is entitled to buy **another** two at a discount of $0.30 off the original price each after buying three.

Owen has $24. Find the amount of money he will have in the end after buying the maximum number of ice cream cones.

- A. $0.50
- B. $0.30
- C. $0.40
- D. $0.20

Question: 29 of 35

Roads A and B were of the same length. Each of the 21 street lamps on Road A was 10 m apart. Road B had 4 fewer street lamps than Road A. Find the distance between 2 street lamps on Road B.

- A. 14.5m
- B. 16.5 m
- C. 15.5 m
- D. 12.5 m

Question: 30 of 35

ABCD is a square piece of paper. A corner of the paper was folded to form triangle EFG. Find ∠y.

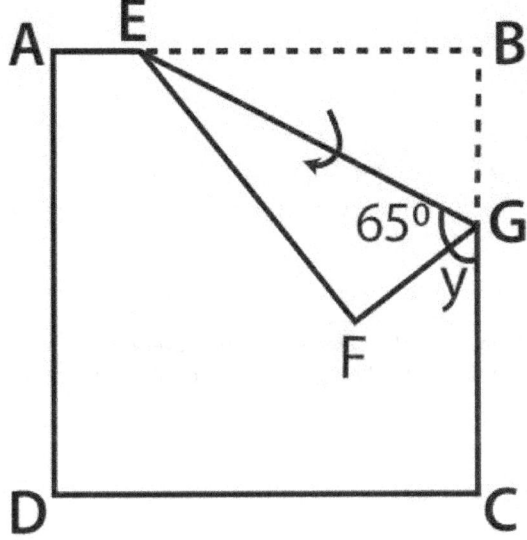

- A. 60
- B. 65
- C. 50
- D. 55

Question: 31 of 35

Lula is selling cakes to raise money for charity.
The graph shows how the amount of money left to raise is related to the number of cakes that Lula has sold.

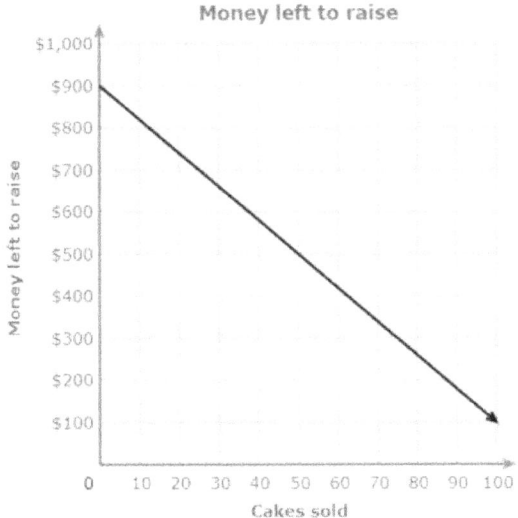

Guess the number of cakes Lula needs to sell in order to have just $500 left to raise.

- A. 48
- B. 37
- C. 50
- D. 30

Question: 32 of 35

Which sign makes the statement true?

$\frac{8}{9}$? $\frac{4}{9} + \frac{1}{3}$

- A. >
- B. <
- C. =
- D. -

Question: 33 of 35

Alicia read 24 pages of a story book every day while Brenda read 16 pages of the same story book every day. Brenda started reading the book on Monday, 2 days ahead of Alicia. On which day would both of them be on the same page?

- A. Wednesday
- B. Thursday
- C. Friday
- D. Saturday

Question: 34 of 35

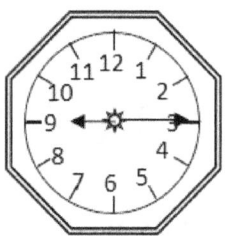

Henry wakes up and looks at the clock. He has 20 minutes to get ready for school. However, he got ready in 10 minutes and rode his bike to the school. It took 15 minutes to reach the school gates and 5 minutes to reach the class.
At what time does Henry arrive at the class?

- A. 9:45 a.m.
- B. 9:50 a.m.
- C. 9:35 a.m.
- D. 9:55 a.m.

Question: 35 of 35

Diagram below is an incomplete bill that shows the items bought by Alice.

Quantity	Item	Price per kilogram	Price
1 kg	Grape	$17	
4 kg	Guava		
		Total	$35

Alice went back to the store and bought 7kg more guavas. Calculate the price for all the guavas that Alice bought.

- A. $50
- B. $55.50
- C. $49.50
- D. $49

Preptive Prepare Better
Series J Thinking Skill Trial Test

Name: _____

Date: _____

Question: 1 of 40

Tom and Harry are shooting marbles. The winner of each round collects a third of the marbles of the loser. Both started the game with 30 and 45 marbles respectively.
If Tom won the first round and Harry the second round, how many marbles would Tom have at the beginning of the third round?

- A. 30
- B. 42
- C. 45
- D. 50

Question: 2 of 40

A bookstore has 100 fiction novels and 50 non-fiction novels in stock. Each of the fiction novels sells for a price between $5 and $8, and each of the non-fiction novels sells for a price between $10 and $14.

Is the average price of all the novels in stock at the bookstore lower than $6?

- A. Yes, lower than $6
- B. No, higher than $6
- C. Same as $6
- D. Can not be determined

Question: 3 of 40

It is a boring day, Jerry is forced to stay indoors. So he decided to read. In his dad's library are some novels of different genres. These included 26 mysteries, 12 modern romances, 12 science fiction, 8 classical romances, and 10 general novels.

How many novels maximum can he read if half of the mysteries and a third of the science fiction that are his father's favourites are out of bound to him?

- A. 47
- B. 660
- C. 51
- D. 48

Question: 4 of 40

A Social Science Club is to attend a quiz competition with 3 representatives participating in the 3 subject areas. The club decided to hold a mock quiz among its members to pick the team to represent it at the competition.
The following are the results best performances of the mock quiz.

	Member	Scores in percentage (equal importance)		
		Politics	Philosophy	Economics
1	Rudolph	87	92	90
2	Betty	80	95	76
3	Pamela	91	90	90
4	Sheila	88	89	83
5	Juliet	70	83	85
6	Garry	60	91	82
7	Sam	92	97	96
8	Lucas	73	92	75

The club is to choose a 3-person team to represent it in the competition, based on the performance at the mock. Average score across subjects count.

Who are the best members of the team?

- A. Sam, Pamela, Betty
- B. Sam, Betty and Pamela
- C. Sam, Rudolph and Pamela
- D. Sam, Betty and Rudolph

Question: 5 of 40

The opinion polls show that Mr McDonald of DDP is leading in the gubernatorial election poll by 45% to 35% over Mr Bailey of PCE. 20% of those polled are undecided.

Tim: "Mr McDonalds is set to be the next Governor."

Which one of the following sentences shows the mistake in Tim's statement?

- A. Mr McDonald is leading his opponent in the opinion polls.
- B. The margin of undecided is large enough for Mr Bailey to possibly win the election
- C. Mr McDonald is more popular among those polled.
- D. Mr Bailey is 10 points behind Mr McDonald in the polls

Question: 6 of 40

Bill scored 86% on a Science test. And he performed better than Graham. Joe led the class in the test while Ted scored the least.
Given the information above, which one of the following **cannot be true**?

- A. Graham scored better than Ted
- B. Joe scored above 80 %
- C. Ted was placed between Bill and Graham
- D. Billy was placed between Joe and Graham

Question: 7 of 40

Four friends: Boyle, Ted, Henry and Dylan are members of a soccer team. They are goalkeeper, midfielder, striker, and defender but not in that order.
Two are left-footers. One of the left-footers is a goalkeeper, the other is a midfielder
Boyle, who is a striker, scores with his right foot as the dominant one
Ted is a left-footer and outfield player (meaning not goalkeeper).

Who is the other right footer?

- A. A defender
- B. A winger
- C. A Goalkeeper
- D. A midfielder

Question: 8 of 40

Shipping a cargo from one town to another involves taking some decisions. The consignment can be shipped either through the train or truck or by sea. The costs and capacity of the available transportation are as follows:

Means of Transport	Capacity	Cost of rental
Road	9kg	$28
Rail	15kg	$50
Sea	20kg	$60

If the weight of the consignment is 360kg, which means should be used?

- A. Road
- B. Rail
- C. Sea
- D. All are same

Question: 9 of 40

Statement : "If the floods don't end soon, most students will struggle to report to school due to inaccessibility. This is a big problem."

Jack: "Most students hail from areas affected by floods."
Lucy: "Accessibility to school is important to students."

Which of the following reasoning is correct?

- A. Jack Only
- B. Lucy Only
- C. Both Jack and Lucy
- D. Neither Jack nor Lucy

Question: 10 of 40

The rules of a certain language are as follows.

- Any word that starts with a consonant has its first letter replaced by !
- A word that starts with a vowel has its first letter replaced by @.
- A word that starts and ends with a consonant has the first and the last letters replaced by # and $, respectively.
- A word that starts and ends with a vowel has the first and last replaced by % and ^, respectively.

Which of the following is a possible code for the word ARROW?

- A. @1234
- B. !1234
- C. #123$
- D. %123^

Question: 11 of 40

A new energy drink called "EnerGize" has been launched recently, claiming to boost energy levels by 200% in just 15 minutes. A group of university students decided to test its effectiveness. After consuming the drink, they felt more energetic during their study session. Based on this, Sarah concluded, "EnerGize can enhance anyone's energy levels by 200% within 15 minutes."

Which statement below best identifies a flaw in Sarah's conclusion?

- A. Sarah's conclusion is based on a small sample size of university students and may not represent the wider population's response to the energy drink.
- B. EnerGize's claim about boosting energy by 200% is unrealistic and scientifically impossible.
- C. Sarah's friends didn't feel any effects after consuming the energy drink, so it must not work for everyone.
- D. Sarah's assumption is correct, as she and her friends experienced increased energy levels after consuming EnerGize.

Question: 12 of 40

If Scarlet is the eldest in her family and is now 25 years old, which of the following statements is also true if none of her siblings share the same birth year and the year of reference is 2022?

- A. The third born hadn't been born in 2000
- B. No one among the siblings was born in 1995
- C. The family has twins
- D. The last born is at least 20 years old

Question: 13 of 40

A library organizes its books using a unique code system. Each code consists of a letter followed by three numbers. The numbers wrap around 999. The letter represents the genre (F for fiction, N for nonfiction, H for history), and the numbers represent the book's position on the shelf.
If the code "N521" is distantly followed by "N003" on the shelf, which of the following codes could be missing between them?

- A. N499
- B. H105
- C. N522
- D. F520

Question: 14 of 40

A mysterious machine operates under these rules:

- It accepts words as input.
- It reverses the order of letters in the input word.
- If a vowel appears in the reversed word, it repeats that vowel at the end of the output.
- If no vowels appear in the reversed word, it adds "-tion" to the end of the output.

What would be the machine's output for the input "strange"?

- A. egnarts
- B. egnartsae
- C. egnart
- D. egnartion

Question: 15 of 40

On a journey through the mystical land of Aethel, you encounter three enigmatic strangers: the Oracle, the Seer, and the Whisperer. You know:

- The Oracle always tells the truth.
- The Seer always lies.
- The Whisperer alternates between truth and falsehood with each statement.

You ask each stranger the same question: "Which of us is currently on the path to enlightenment?"

The Oracle says, "The Seer is enlightened."
The Seer says, "The Whisperer is not enlightened."
The Whisperer says, "The Oracle is lying."

Based on their pronouncements, who among the three may not be on the path to enlightenment?

- A. The Oracle
- B. The Seer
- C. The Whisperer
- D. All three could be on the path.

Question: 16 of 40

Professor Elara, a renowned historian, investigates the reign of Queen Anya. She discovers:

- All of Queen Anya's loyal advisors were also skilled swordsmen.
- No skilled swordsman would ever betray their queen.
- Queen Anya's advisor Sir Gareth was not a skilled swordsman.

Can Professor Elara conclude with certainty that Sir Gareth betrayed Queen Anya?

- A. Yes, the conclusion is logically valid.
- B. Yes, but only with additional evidence.
- C. No, the conclusion is logically invalid.
- D. The conclusion is irrelevant to the investigation.

Question: 17 of 40

In the labyrinthine Library of Eldoria, each book contains a single true statement and a single false statement.

The librarian claims:
- If the book you are holding is true, then the book to the right is false.
- If the book you are holding is false, then the book to the left is true.

You pick up a book at random. Based on the librarian's claim, what can you definitely conclude about the book in your hand?

- A. It is true.
- B. It is false.
- C. It is either true or false, but we cannot know which.
- D. The statement about the librarian's claim is itself untrue.

Question: 18 of 40

A gardening expert advises, "Rotating crops in your garden helps maintain soil fertility."

Which statement, if true, best supports the expert's advice?

- A. Some gardeners believe that using synthetic fertilizers eliminates the need for crop rotation.
- B. Limited space in urban gardens makes it challenging for individuals to rotate crops effectively.
- C. Modern irrigation systems can compensate for soil nutrient depletion, reducing the importance of crop rotation.
- D. Crop rotation prevents the depletion of specific nutrients in the soil, promoting long-term soil fertility.

Question: 19 of 40

Since the winters will be extremely cold this year, Mike and Melody won't be going for their vacation in Spain. Instead, they plan to go to a new local restaurant. It means spending less than the initial budget. Unless they find availability, they won't spend the extra money in case they decide to go to Spain later.

If these are facts, which of the following statements is also true?

- A. Mike and Melody have entirely given up on visiting Spain.
- B. It is now final that Mike and Melody will be going to the new local restaurant.
- C. The cold winters have affected Mike and Melody's decision to visit Spain.
- D. A trip to Spain is overpriced.

Question: 20 of 40

Cosmos is an American sports club founded in 1976 by a sports enthusiast, Ferd Ackerman and his group of wealthy friends. They first formed a Baseball team, The Hard-hitters, in the group's home town, Detroit. Then, the group organized a Basketball team, the Jumpers.
The group first participated in American Football with the formation of a team in Cincinnati, the Cincinnati Kickers, in 1989. At that time the team was formed to participate in the sport's inner town league in the city. It was an innovation to inner town team sports and the new game of American Football. The group also formed the Jersey Dodgers to participate in the sport's Eastern Conference League. The San Francisco Smashers, its most successful team to date, was formed in 1992 to participate in the new sport's California State League. This team has gone on to win the State Championship 12 times since its formation.
After this club, came the formation of the Atlanta Cranes in 2008. After this, the group's appetite for new teams seems to have waned as no new club has been formed since.

Which sports team is the San Francisco Smashers?

- A. Basketball
- B. Baseball
- C. American football
- D. Squash

Question: 21 of 40

This device should always be repaired by a professionally trained technician and handled carefully at all times.
Children must not be allowed to operate this device nor anybody under the influence of alcohol.
Repairs to this equipment may only be carried out by trained professionals or at the manufacturer's repair centres across the country.
Improper or unauthorized repairs can lead to considerable risks to users, including electrical shocks.

To which appliance would this user manual belong?

- A. A water heater
- B. A bookshelf
- C. A toy box
- D. An android tablet

Question: 22 of 40

Every day Brian walks to school from his house, which is 1.4km away. However, on Saturdays, he goes to his father's workshop to help out before going for his football training in school; and then comes back home from school. His father's workshop is 700 metres from his home and 400 metres from his school.

How much is the difference between his walks to and from school on weekdays and Saturdays?

- A. 0.2km
- B. 0.3km
- C. 0.4km
- D. 0.56km

Question: 23 of 40

The Grand Prix race at Silverstone started 25 minutes behind schedule because of a snow blizzard. The race was won by the Ferrari driver with a time of one hour and twelve minutes. The second-placed driver in the Mercedes car, came in 25 minutes after the Ferrari car crossed the line and the third-placed driver came in six minutes after the Mercedes car. At what time did the third-placed driver cross the line if the race was originally scheduled to begin at 3.50 pm?

- A. 5:58 pm
- B. 4:33 pm
- C. 4:56 pm
- D. 4:34 pm

Question: 24 of 40

The Red Devils have won the League more than most teams in the League, even though they have not won the title in the last three years as The Saints have won consecutively in the last two years.

Tim: "The Red Devils are the best team in the League, having won the title 7 times in the last 10 years."
Patrick: "The Saints are the current Champions in the league."

If the information is true, whose reasoning is correct?

- A. Tim only
- B. Patrick only
- C. Both are wrong
- D. Both are right

Question: 25 of 40

Randall Kendham was a great middle-distance runner who first broke into national consciousness as an athlete at the University of Olympio. Then he broke the 1,500m National Records, which also was the National Collegiate Records. While the national record was broken nine years later, the Collegiate record of 3:40 mins stood for a long time until it was broken last year by an athlete from the University of Dolphin, who ran a semi-final race of the 2022 Collegiate Games at 3:29 mins.

What fraction of the original record time did the new record holders take off ?

- A. $3/10$
- B. $6/7$
- C. $1/20$
- D. $1/30$

Question: 26 of 40

There are three diners in a restaurant, having late meals. Two are regular and the other is just a casual customer at the restaurant. One had a meal of French fries, another hamburger and the third mashed potatoes.
The regulars decided to take the house coffee - one Spanish the other Irish as desserts, while the third man took ice cream.

After the meals, the man who ate fries left a large tip, the man who ate potatoes left a normal tip after his cup of Spanish coffee and the third man left no tip after his ice cream.

What did the man with the large tip drink?

- A. Spanish Coffee
- B. Irish Coffee
- C. Ice Cream
- D. Water

Question: 27 of 40

In a league competition, the 10 teams have to play each other at home and away. With three matches left, *The Park Rangers* sit atop the table with 51 points, the next-placed team has 43 points and the third-placed team has 41 points.

If a win attracts 3 points, a draw 1 point and a loss no point, what is the least possible position Park Ranger could finish the season with?

- A. First
- B. Second
- C. Third
- D. Fourth

Question: 28 of 40

In a class of 30 students, Michael was in the top half of the class, Iris was in the top third of the class, Peggy's position was 16th and Ramos was among the first 5 brilliant students in the class.

How are these students placed in class, if all students are near end in their category?

- A. Michael, Iris, Peggy, Ramos
- B. Ramos, Michael, Iris, Peggy
- C. Ramos, Iris, Michael. Peggy
- D. Michael, Ramos, Iris, Michael

Question: 29 of 40

The chef uses only tender boneless meat to cook the special dish. There are 28 pieces of meat in the freezer. 12 are tough meat and the **rest** are tender.

If 7 of the **rest** are bony, how many pieces of the meat can be used for the special dish?

- A. 10
- B. 12
- C. 8
- D. 9

Question: 30 of 40

The year is made up of 4 seasons: Spring, Summer, Fall and Winter, each with different weather, temperature and climatic conditions.

Dick: "Each year, the world undergoes four climatic changes."
Peter: "There are different weathers in a year."
Tom: "Some of the seasons' weathers are similar to each other."

If the information is true, whose **inference** is correct?

- A. Dick and Peter only
- B. Peter and Tom only
- C. Dick and Tom only
- D. None of the options provided

Question: 31 of 40

The following is the number of hats owned by 3 gentlemen.

- Leslie: black hats (4); white hats (1); brown hats (1)
- Patrick: black hats (2); white hats (2) brown hats (2)
- Arthur: black hats (4); white hats (1); brown hats (3)

If the value of Arthur's hats is worth $30 more than that of Leslie's, and a Black hat sells for $15 and a white hat sells for $20, what is the difference between the value of Patrick's hats and Arthur's?

- A. $45
- B. $35
- C. $40
- D. $25

Question: 32 of 40

Four friends, Smith, Beatrice, Fred, and Diana, each went to see a different movie: action drama, comedy, family drama, and thriller. You know that:
- Smith and Beatrice saw movies in somewhat similar genre.
- Fred did not see the comedy movie.
- Diana saw either the comedy or the thriller.

If Smith saw the family drama movie, what movie did Diana see?

- A. Action
- B. Comedy
- C. Drama
- D. Thriller

Question: 33 of 40

A pirate map leads to a buried treasure on a deserted island. The map has four instructions:
- Turn right and walk for 5 paces.
- Turn right again and walk for 15 paces.
- Dig at the spot.

If you start facing east, where will you be digging for the treasure?

- A. North
- B. East
- C. South-West
- D. North-West

Question: 34 of 40

Three friends, Joy, Noah, and Olivia, each brought a different fruit to share at a picnic: apple, banana, and orange. You know that:
- Joy likes both apples and bananas.
- Noah does not like oranges.
- Olivia brought the fruit that Joy likes.

If Noah brought the apple, what fruit did Olivia bring?

- A. Apple
- B. Banana
- C. Orange
- D. Impossible to determine

Question: 35 of 40

Three friends—Oliver, Penelope, and Quincy—embark on a quest through the Enchanted Forest. The forest has magical creatures and hidden paths with the following guiding principles:

- Principle A: "The path with fireflies is always taken before passing through the dense thicket."
- Principle B: "Once the crystal-clear pond is found, the journey must continue without encountering the mystical wolves."
- Principle C: "Talking owls are encountered immediately after passing through the dense thicket."

If Quincy encounters the talking owls, what can be inferred about the previous elements of the journey?

- A. The fireflies were encountered.
- B. The crystal-clear pond was found.
- C. The mystical wolves were bypassed.
- D. Impossible to determine

Question: 36 of 40

In the heart of the enchanted forest, there are three magical creatures: Gryphor, Mystique, and Zephyr. Each creature guards a different path leading to a hidden realm. The inscriptions near their lairs read:

- Gryphor: "Choose the path of courage, where the mightiest trees stand tall."
- Mystique: "Navigate through the shadows, where illusions conceal the way."
- Zephyr: "Follow the direction of the wind, where the air hums the secrets of the realm."

You have a compass and weather app on your smartphone. If you seek the hidden realm, which path should you choose?

- A. Gryphor's path
- B. Mystique's path
- C. Zephyr's path
- D. Impossible to determine

Question: 37 of 40

Which of the following is the correct figure?

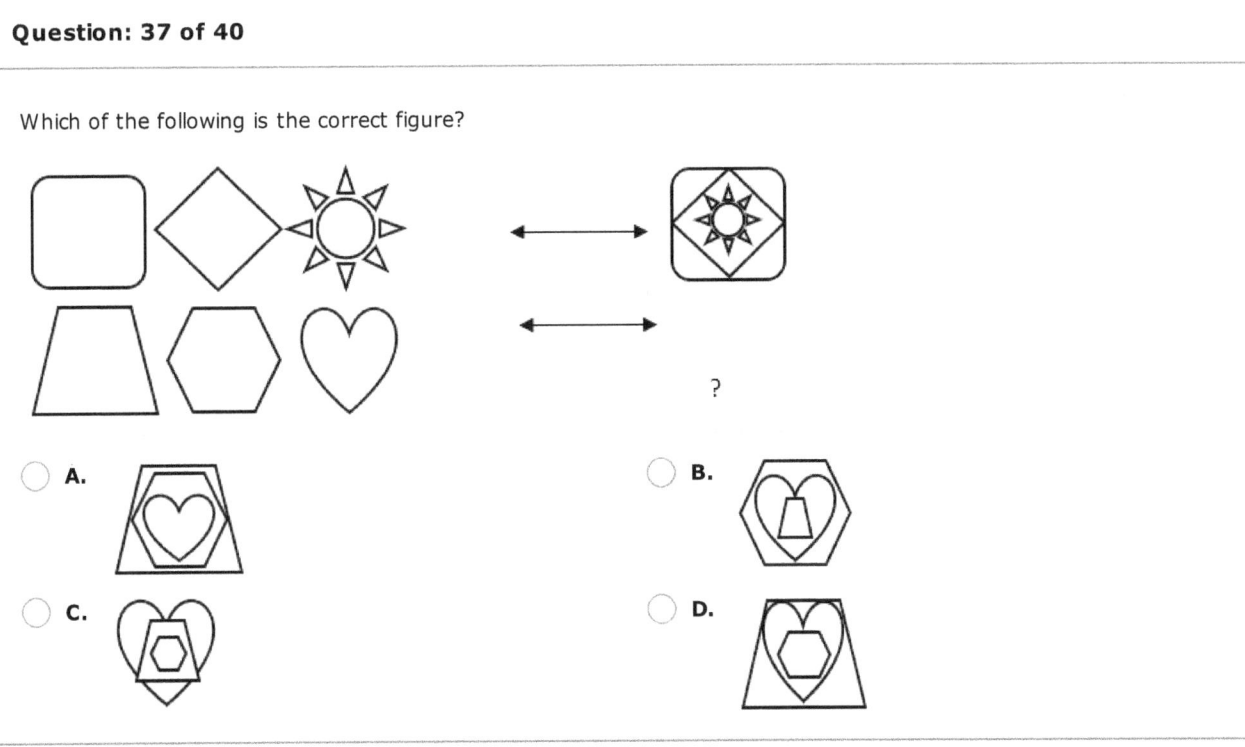

Question: 38 of 40

In water, a word is shown as ⟨mirror image: PRAKASH⟩.
What is correct form of the word?

A. PRASHAK
B. PRAKASH
C. PRSKHAP
D. KRASHAP

Question: 39 of 40

Select the figure from the options which when placed in blank space would complete the pattern.

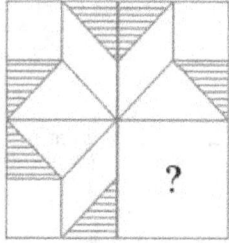

A.
B.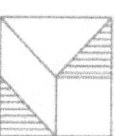
C.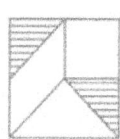
D.

Question: 40 of 40

Find the values of P and Q respectively in the given number pattern.

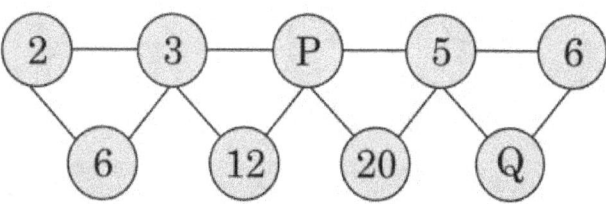

A. 5 and 28
B. 4 and 30
C. 3 and 30
D. 2 and 18

Preptive Prepare Better
Series K Reading Comprehension Trial Test

Name: _____

Date: _____

Direction for Questions: 1 to 5

Read the two extracts and then answer the questions.

Extract A: The Titanic: A Monument to Human Ambition and Tragedy

The RMS Titanic, heralded as the "unsinkable" marvel of its time, remains one of the most iconic and tragic maritime disasters in history. This majestic ocean liner, touted as a testament to human ingenuity and ambition, embarked on its maiden voyage with great fanfare, only to meet a fate that would echo through the annals of time. As we delve into the story of the Titanic, we uncover not only the grandeur of its design but also the harrowing tale of human frailty and resilience that unfolded on that fateful night.

At the dawn of the 20th century, the Titanic stood as a symbol of modernity and luxury, boasting unparalleled opulence and technological advancements. Constructed by the renowned shipbuilding company Harland and Wolff in Belfast, Ireland, the Titanic was the largest and most luxurious ship of its time, designed to ferry passengers across the Atlantic with unparalleled comfort and style. From its exquisite dining halls and lavish cabins to its state-of-the-art amenities, the Titanic represented the epitome of luxury travel, attracting passengers from all walks of life, from the wealthiest elite to the humblest immigrants seeking a new beginning in the land of opportunity.

However, beneath its veneer of opulence lay a fatal flaw that would seal the Titanic's tragic fate. On the fateful night of April 14, 1912, the Titanic collided with an iceberg in the frigid waters of the North Atlantic, setting into motion a chain of events that would culminate in one of the deadliest maritime disasters in history. Despite the ship's advanced safety features, including watertight compartments and an extensive network of lifeboats, the sheer magnitude of the collision proved catastrophic, leading to the rapid flooding of multiple compartments and the eventual sinking of the ship.

The ensuing chaos and desperation that unfolded on board the Titanic revealed both the best and worst of human nature. As the ship began to sink into the icy depths, acts of heroism and selflessness emerged amidst the panic and despair. From the gallant efforts of the crew to evacuate passengers to the poignant sacrifices made by individuals who gave up their own lives to save others, the tragedy of the Titanic showcased the resilience and courage of the human spirit in the face of adversity.

Yet, amidst the tales of heroism and survival, the sinking of the Titanic also exposed the glaring shortcomings and oversights that contributed to the scale of the disaster. Despite warnings of icebergs in the vicinity and the lack of a sufficient number of lifeboats to accommodate all passengers and crew, complacency and hubris prevailed, leading to a catastrophic loss of life that could have been prevented. The sinking of the Titanic served as a sobering reminder of the consequences of human arrogance and the importance of humility and vigilance in the face of nature's unpredictability.

In the aftermath of the tragedy, investigations were launched, inquiries were held, and reforms were enacted to prevent such a disaster from ever happening again. The sinking of the Titanic prompted sweeping changes in maritime safety regulations, including the implementation of stricter standards for ship design, navigation procedures, and the provision of life-saving equipment. The lessons learned from the Titanic disaster continue to shape maritime safety practices to this day, ensuring that future generations of seafarers remain vigilant and prepared to confront the challenges of the open ocean.

As we reflect on the legacy of the Titanic, we are reminded of the fragility of human existence and the enduring allure of the sea. The Titanic stands not only as a monument to human ambition and technological prowess but also as a solemn memorial to the lives lost and the lessons learned from one of the most tragic chapters in maritime history. May we never forget the sacrifices made and the lives lost aboard the Titanic, and may their memory serve as a beacon of remembrance and reverence for generations to come.

Extract B: What History Did Not Recognize About Titanic: Unveiling Untold Stories

The sinking of the RMS Titanic is a story etched into the annals of history, a tale of tragedy and heroism that has captivated the collective imagination for over a century. Yet, amidst the well-documented accounts and widely known narratives, there exist untold stories and overlooked aspects of the Titanic saga that offer new perspectives and deeper insights into this monumental event. As we peel back the layers of history, we uncover the hidden truths and lesser-known facets of the Titanic's legacy.

One of the lesser-known aspects of the Titanic story is the role played by the ship's passengers and crew in shaping the events that unfolded on that fateful night. While the stories of the wealthy elite and prominent figures aboard the Titanic have been widely recounted, the experiences of the ordinary men, women, and children who made up the majority of the ship's complement often go unnoticed. From the immigrant families seeking a better life in America to the crew members who worked tirelessly behind the scenes, each individual aboard the Titanic had a unique story to tell, offering a mosaic of perspectives that enrich our understanding of the human drama that unfolded amidst the chaos and despair of the sinking ship.

Another overlooked aspect of the Titanic saga is the impact of the disaster on the communities and families left behind. Beyond the headlines and sensationalized accounts, the sinking of the Titanic reverberated across continents, leaving a trail of grief and loss in its wake. From the small towns of Ireland and England, where entire communities mourned the loss of loved ones, to the shores of America, where immigrant families awaited news of their relatives' fate, the tragedy of the Titanic transcended national boundaries, leaving an indelible mark on the collective consciousness of humanity

Furthermore, the aftermath of the Titanic disaster had far-reaching consequences that extended beyond the immediate loss of life. The sinking of the Titanic prompted significant reforms in maritime safety regulations, including the implementation of stricter standards for lifeboat capacity, navigation practices, and wireless communication procedures. These reforms were instrumental in improving the safety of future ocean voyages and preventing similar disasters from occurring.

In addition to its historical significance, the Titanic has also become a symbol of human curiosity and exploration. Since its discovery in 1985, the wreckage of the Titanic has been the subject of numerous scientific expeditions and research endeavors aimed at unraveling the mysteries surrounding its final moments. Through advanced imaging technology and underwater exploration, researchers have been able to piece together a more comprehensive understanding of the events that led to the Titanic's sinking and the conditions of the wreck site.

Yet, despite our efforts to uncover the truth about the Titanic, there are still unanswered questions and unresolved mysteries surrounding the disaster. The precise sequence of events leading up to the collision with the iceberg, the actions of the crew in the moments before the sinking, and the experiences of individual passengers and crew members remain subjects of speculation and debate.

In conclusion, the story of the Titanic is not just a tale of a tragic maritime disaster but a multifaceted narrative that encompasses themes of social inequality, human resilience, and the pursuit of knowledge. By exploring the lesser-known aspects of the Titanic's history, we gain a deeper appreciation for the complexities of the event and the enduring legacy of one of the most iconic disasters in modern history.

Question: 1 of 30

What does Extract B reveal about the passengers and crew of the Titanic?

- **A.** Their collective disregard for safety protocols
- **B.** Their diversity in social and economic backgrounds
- **C.** Their advanced knowledge of maritime navigation
- **D.** Their unanimous decision to abandon ship

Question: 2 of 30

What role do unanswered questions play in the narrative of the Titanic disaster, as discussed in Extract B?

- **A.** They serve as evidence of conspiracy theories.
- **B.** They fuel ongoing debates among historians.
- **C.** They detract from the significance of the event.
- **D.** They are dismissed as irrelevant to understanding the disaster.

Question: 3 of 30

How does Extract A characterize the sinking of the Titanic in terms of human nature?

- **A.** As a display of unwavering courage and resilience
- **B.** As a reflection of human arrogance and humility
- **C.** As a testament to human greed and corruption
- **D.** As a result of external factors beyond human control

Question: 4 of 30

What aspect of the Titanic's legacy does Extract B primarily seek to explore?

- A. Its lasting impact on maritime technology
- B. Its influence on popular culture and media
- C. Its untold stories and lesser-known facets
- D. Its role in shaping historical narratives

Question: 5 of 30

According to Extract A, what major reforms were implemented in response to the Titanic disaster?

- A. Stricter regulations on passenger accommodations
- B. Enhanced training for maritime crews
- C. Improved communication systems and navigation procedures
- D. Increased reliance on luxury amenities

Direction for Questions: 6 to 9

Read the two extracts and then answer the questions.

Extract A: Food Security: Nourishing Nations for Sustainable Futures

Food security stands as a cornerstone of societal well-being, encompassing the availability, access, and utilization of nutritious food for all individuals within a population. In a world grappling with diverse challenges, from climate change and population growth to socioeconomic disparities, ensuring food security emerges as a pressing imperative for fostering resilient and sustainable communities.

At its essence, food security entails more than just the provision of calories; it encompasses the availability of diverse and nutritious foods that meet the dietary needs of individuals across all age groups. Achieving food security requires a multifaceted approach that addresses various dimensions, including agricultural productivity, food distribution systems, economic access, and nutritional education.

Central to the concept of food security is the notion of resilience – the capacity of food systems to withstand shocks and stresses while maintaining adequate levels of food availability and access. Climate change, characterized by extreme weather events, shifting precipitation patterns, and rising temperatures, poses a significant threat to agricultural productivity and food security worldwide. Adapting agricultural practices to mitigate the impact of climate change, promoting drought-resistant crops, and investing in sustainable water management strategies are crucial steps towards building resilient food systems.

Equally critical is addressing the structural inequalities that underpin food insecurity, particularly in marginalized communities. Socioeconomic factors, including poverty, unemployment, and inadequate access to education and healthcare, exacerbate vulnerability to food insecurity. Empowering communities through livelihood diversification, social protection programs, and equitable access to resources can enhance their capacity to secure an adequate and nutritious diet.

Moreover, the globalization of food systems has profound implications for food security, shaping patterns of production, trade, and consumption on a global scale. While international trade can enhance food access and diversity, it also exposes countries to market volatility and supply chain disruptions. Strengthening local food systems, promoting agroecological practices, and fostering food sovereignty empower communities to assert greater control over their food sources and reduce dependency on volatile global markets.

In the pursuit of food security, nutrition emerges as a central tenet, recognizing the intrinsic link between food consumption and health outcomes. Malnutrition, encompassing undernutrition, micronutrient deficiencies, and overweight/obesity, remains a persistent challenge globally, contributing to a double burden of malnutrition in many regions. Promoting diverse and balanced diets, improving access to clean water and sanitation, and enhancing nutrition education are essential strategies for addressing malnutrition and promoting holistic well-being.

In conclusion, food security stands as a fundamental human right and a cornerstone of sustainable development. Achieving food security requires a comprehensive and integrated approach that addresses the complex interplay of environmental, social, economic, and health factors. By fostering resilient food systems, promoting equity and social justice, and prioritizing nutrition and health, societies can nourish their populations and build a foundation for thriving and sustainable futures.

Extract B: Why Most African Countries Face Food Insecurity: Unraveling the Complexities

Food insecurity, a persistent challenge faced by many African countries, stems from a complex interplay of factors that encompass environmental, social, economic, and political dimensions. Despite the continent's rich natural resources and agricultural potential, widespread poverty, climate variability, inadequate infrastructure, and governance challenges contribute to the prevalence of food insecurity in many regions. Let's delve into the multifaceted reasons why most African countries grapple with food insecurity.

1. Climate Variability and Environmental Degradation: African countries are particularly vulnerable to climate variability and extreme weather events, such as droughts, floods, and cyclones. These climatic shocks disrupt agricultural production, leading to crop failures, livestock losses, and diminished food availability. Environmental degradation, including deforestation, soil erosion, and land degradation, further exacerbates vulnerability to food insecurity, undermining the resilience of rural livelihoods.
2. Limited Access to Resources and Technology: Many smallholder farmers in Africa lack access to essential resources, including land, water, seeds, and agricultural inputs. Limited access to modern farming technologies, such as mechanization, irrigation, and improved seeds, constrains agricultural productivity and exacerbates food insecurity. Furthermore, inadequate infrastructure, including roads, storage facilities, and market access, hinders the efficient distribution of food and agricultural products.
3. Poverty and Economic Inequity: Persistent poverty and economic inequality are significant drivers of food insecurity in Africa. High levels of unemployment, low wages, and limited access to financial services trap millions of people in poverty, limiting their ability to afford an adequate and nutritious diet. Economic shocks, such as currency devaluation and inflation, further erode purchasing power, exacerbating food insecurity among vulnerable populations.
4. Conflict and Political Instability: Conflict and political instability are key drivers of food insecurity in several African countries. Armed conflicts, civil unrest, and displacement disrupt food production, distribution, and access, leading to food shortages and humanitarian crises. In conflict-affected regions, food aid delivery is often hampered by logistical challenges and security risks, exacerbating the plight of affected populations.
5. Population Growth and Urbanization: Rapid population growth and urbanization place significant pressure on food systems in many African countries. Urbanization drives demand for food, while simultaneously reducing the availability of arable land and rural labor. As a result, many African countries rely heavily on food imports to meet the dietary needs of their growing urban populations, increasing vulnerability to global market fluctuations and price volatility.
6. Governance and Policy Challenges: Weak governance, corruption, and inadequate policy frameworks hinder efforts to address food insecurity in many African countries. Inefficient agricultural policies, lack of investment in rural development, and limited support for smallholder farmers constrain agricultural productivity and food production. Moreover, inadequate social safety nets and emergency response mechanisms leave populations vulnerable to food crises and humanitarian emergencies.

In conclusion, addressing food insecurity in Africa requires a holistic approach that addresses the underlying drivers and root causes of vulnerability. Investing in sustainable agricultural practices, improving access to resources and technology, strengthening social safety nets, and promoting good governance are essential steps towards achieving food security and fostering resilient and thriving societies across the continent.

Question: 6 of 30

According to Extract B, what is a significant challenge faced by African countries due to rapid urbanization?

- A. Decreased demand for food
- B. Increased availability of arable land
- C. Higher dependence on food imports
- D. Improved access to agricultural resources

Question: 7 of 30

Which concept is central to the discussion of food security in both extracts?

- A. Economic inequality
- B. Malnutrition
- C. Resilience
- D. Urbanization

Question: 8 of 30

According to both extracts, what is critical for fostering resilient and sustainable communities?

- A. Increased food imports
- B. Promotion of monoculture farming
- C. Investment in rural development
- D. Dependence on global markets

Question: 9 of 30

How does Extract B characterize the impact of conflict and political instability on food security?

- A. They enhance food production and distribution.
- B. They have minimal effect on food availability.
- C. They disrupt food production and access.
- D. They facilitate efficient food aid delivery.

Direction for Questions: 10 to 14

Read the extracts and then answer the questions.

Extract A: Metal roofing stands as a modern marvel in the construction world. Known for its durability and sleek appearance, metal roofs have gained popularity among homeowners and architects alike. These roofs, often crafted from steel, aluminum, or copper, offer exceptional longevity and resistance to harsh weather conditions. Their reflective properties also contribute to energy efficiency, reducing cooling costs during scorching summers. Additionally, metal roofs come in various styles, from standing seam to metal tiles, catering to diverse aesthetic preferences.

Extract B: Slate roofing embodies timeless elegance and durability. Quarried from natural stone, slate tiles boast a unique, sophisticated appeal that enhances the aesthetics of any structure. The distinct textures and colors available in slate allow homeowners to create visually striking roofs. Beyond their visual appeal, slate tiles are renowned for their longevity, with some roofs lasting over a century. While heavier and more labor-intensive to install, the resilience and natural beauty of slate make it a cherished choice for those seeking enduring sophistication.

Extract C: Asphalt shingles remain a popular and cost-effective choice for residential roofing. These shingles, made from fiberglass or organic materials coated with asphalt, provide a diverse range of colors and styles, catering to various architectural designs. Their ease of installation and affordability make them a go-to option for many homeowners. Additionally, asphalt shingles offer reliable protection against moderate weather conditions and come with warranties that ensure their durability for several decades, making them a practical choice for those seeking a balance between cost and performance.

Extract D: Green roofing emerges as an innovative and environmentally conscious roofing solution. These roofs, adorned with living vegetation, offer a unique blend of sustainability and aesthetic appeal. By incorporating vegetation and specialized layers, green roofs provide natural insulation, reducing energy consumption and mitigating the urban heat island effect. They contribute to air purification, stormwater management, and biodiversity in urban areas. While requiring specific structural considerations and maintenance, green roofs serve as a testament to sustainable living and ecological responsibility in modern construction practices.

Question: 10 of 30

Which extract highlights roofing known for its sophisticated appearance crafted from natural stone?

- A. Extract A
- B. Extract B
- C. Extract C
- D. Extract D

Question: 11 of 30

Which extract mentions roofing that is particularly renowned for its resistance to harsh weather conditions and exceptional longevity?

- **A.** Extract A
- **B.** Extract B
- **C.** Extract C
- **D.** Extract D

Question: 12 of 30

Which extract discusses roofing that offers a balance between cost-effectiveness, diverse styles, and reliable protection against moderate weather conditions?

- **A.** Extract A
- **B.** Extract B
- **C.** Extract C
- **D.** Extract D

Question: 13 of 30

Which extract mentions roofing options that cater to diverse aesthetic preferences through various styles like standing seam and metal tiles?

- **A.** Extract A
- **B.** Extract B
- **C.** Extract C
- **D.** Extract D

Question: 14 of 30

Which extract highlights roofing known for its labor-intensive installation but cherished for its enduring sophistication?

- **A.** Extract A
- **B.** Extract B
- **C.** Extract C
- **D.** Extract D

Direction for Questions: 15 to 18

Read the poem and then answer the questions.

The Chariot
by Emily Dickinson

Because I could not stop for Death,
He kindly stopped for me;
The carriage held but just ourselves
And Immortality.
We slowly drove, he knew no haste,
And I had put away
My labor, and my leisure too,
For his civility.
We passed the school where children played,
Their lessons scarcely done;
We passed the fields of gazing grain,
We passed the setting sun.
We paused before a house that seemed
A swelling of the ground;
The roof was scarcely visible,
The cornice but a mound.
Since then 't is centuries; but each
Feels shorter than the day
I first surmised the horses' heads

Were toward eternity.

Question: 15 of 30

What is the primary theme of Emily Dickinson's poem "The Chariot"?

- A. Celebration of life
- B. Fear of death
- C. Acceptance of mortality
- D. Joy of youth

Question: 16 of 30

What is the significance of the carriage holding "just ourselves And Immortality"?

- A. It suggests the speaker's fear of death
- B. It represents the eternity of the journey
- C. It implies that the journey is lonely
- D. It symbolizes the inevitability of death

Question: 17 of 30

What is the deeper meaning behind the line, "The roof was scarcely visible, The cornice but a mound"?

- A. The speaker is describing a mansion
- B. The house represents a grave
- C. It depicts a house sinking into the ground
- D. It signifies the passing of time

Question: 18 of 30

What emotion does the speaker convey through the line, "Since then 't is centuries; but each / Feels shorter than the day"?

- A. Regret
- B. Fear
- C. Longing
- D. Perceived passage of time

Direction for Questions: 19 to 23

Read the poem and then answer the questions.

Break, Break, Break
By Alfred Tennyson

Break, break, break,
On thy cold gray stones, O sea!
And I would that my tongue could utter
The thoughts that arise in me.
Oh, well for the fisherman's boy,
That he shouts with his sister at play!
Oh, well for the sailor lad,
That he sings in his boat on the bay!
And the stately ships go on
To their haven under the hill;
But oh for the touch of a vanished hand,
And the sound of a voice that is still!
Break, break, break,
At the foot of thy crags, O sea!
But the tender grace of a day that is dead
Will never come back to me.

Question: 19 of 30

What is the primary emotion expressed by the speaker in the poem?

- A. Joy and celebration
- B. Sorrow and longing
- C. Excitement and anticipation
- D. Peace and contentment

Question: 20 of 30

What is the poet's lamentation about in the poem?

- A. The joy of playing by the sea
- B. The absence of fishermen
- C. The loss of his sister's voice
- D. The inability to express his thoughts

Question: 21 of 30

What does the phrase "the touch of a vanished hand" suggest in the poem?

- A. The sensation of cold water
- B. The longing for lost affection
- C. The joy of sailing on the bay
- D. The imagery of a vanished ship

Question: 22 of 30

What is the deeper meaning suggested by the contrast between the activities of the fisherman's boy and the sailor lad?

- A. The speaker's admiration for their adventurous spirit
- B. The speaker's envy of their carefree lives
- C. The juxtaposition of joyous activities against the speaker's sorrow
- D. The speaker's longing for the sea and its delights

Question: 23 of 30

What is the symbolic significance of the sea in the poem?

- A. It represents the speaker's love for sailing
- B. It symbolizes the vastness of nature
- C. It signifies the passage of time and life's impermanence
- D. It embodies the speaker's hope for the future

Direction for Questions: 24 to 30

Seven sentences have been removed from the text. Choose from the sentences (A – H) the one which fits each gap (1 – 7). There is one extra sentence which you do not need to use.

<u>Nurturing Minds and Shaping Futures</u>

The school system is a fundamental institution in society, providing education and shaping the minds of the next generation.**1...** The school system encompasses not only the physical buildings and classrooms but also the curriculum, teachers, administrators, and the overall educational philosophy.

One of the primary purposes of the school system is to impart knowledge and skills to students. It provides a structured environment where students can learn a wide range of subjects, from mathematics and science to history and literature.

2...

In addition to academic learning, the school system also focuses on the development of various skills and abilities. **3...** These skills are essential for success in the modern world, where adaptability and lifelong learning are crucial.

4.... It brings together students from diverse backgrounds and provides them with opportunities to interact, form friendships, and learn from one another. Through extracurricular activities, such as sports, clubs, and organizations, students can explore their interests, develop their talents, and build lasting relationships.

5.... It helps students develop values such as integrity, responsibility, and empathy. Teachers and administrators serve as role models and mentors, guiding students in their journey towards becoming responsible and ethical individuals.

However, it is important to acknowledge that the school system is not without its challenges. **6....** Addressing these inequalities and striving for inclusivity should be a priority for any school system.

7.... On one hand, technology can enhance learning experiences, facilitate access to information, and promote innovative teaching methods. On the other hand, it can also lead to concerns about screen time, privacy, and the potential for increased inequality if not implemented equitably.

In conclusion, the school system is a vital institution that plays a crucial role in nurturing minds and shaping futures. It provides a structured learning environment, promotes socialization and character development, and prepares students for the challenges of the modern world. While there are challenges that need to be addressed, the school system remains an essential foundation for education and society as a whole.

- A. The curriculum is carefully designed to ensure a well-rounded education that equips students with the necessary knowledge to succeed in their academic pursuits and beyond.
- B. Students are encouraged to think critically, solve problems, communicate effectively, and work collaboratively.
- C. Furthermore, the school system serves as a platform for personal and character development
- D. One of the most pressing issues is ensuring equal access to quality education for all students. Disparities in funding, resources, and opportunities can create barriers that prevent some students from reaching their full potential
- E. The school system also plays a vital role in promoting socialization and fostering a sense of community
- F. It provides a structured learning environment, promotes socialization and character development
- G. It plays a crucial role in preparing students for the challenges and opportunities they will face in their lives
- H. The rapid advancement of technology has also brought about both opportunities and challenges for the school system.

Question: 24 of 30

Which one fits in (1) ?

- A. Sentence A
- B. Sentence B
- C. Sentence G
- D. Sentence H

Question: 25 of 30

Which one fits in (2) ?

- A. Sentence A
- B. Sentence B
- C. Sentence C
- D. Sentence D

Question: 26 of 30

Which one fits in (3) ?

- A. Sentence A
- B. Sentence D
- C. Sentence B
- D. Sentence E

Question: 27 of 30

Which one fits in (4) ?

- A. Sentence A
- B. Sentence B
- C. Sentence H
- D. Sentence E

Question: 28 of 30

Which one fits in (5) ?

- A. Sentence A
- B. Sentence B
- C. Sentence C
- D. Sentence D

Question: 29 of 30

Which one fits in (6) ?

- A. Sentence A
- B. Sentence B
- C. Sentence C
- D. Sentence D

Question: 30 of 30

Which one fits in (7) ?

- A. Sentence A
- B. Sentence G
- C. Sentence H
- D. Sentence C

Preptive Prepare Better
Series K Mathematical Reasoning Trial Test

Name: _____

Date: _____

Question: 1 of 35

Joy and Siti had a total of 360 beads at first. Joy lost 28 beads while Siti bought another 18 beads. Both of them had an equal number of beads in the end.

How many beads did **Joy have in the end**?

- A. 157
- B. 175
- C. 185
- D. 203

Question: 2 of 35

The location of three cities A B C is shown in the picture. The distance between A and B is 45 km. The distance between A and C is 161 km.
What is the distance between B and C?

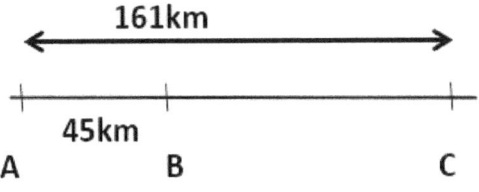

- A. 116km
- B. 45 km
- C. 134km
- D. 125km

Question: 3 of 35

Jin and her brother had some marbles. The ratio of the number of marbles Jin had to the number of marbles that her brother had was 1 : 4 at first. After their mother gave Jin 54 more marbles, Jin and her brother had the same number of marbles.
How many marbles did Jin's brother have at first?

- A. 72 marbles
- B. 18 marbles
- C. 54 marbles
- D. 70 marbles

Question: 4 of 35

Lisa had four times as much money as Joshua. After their mother gave each of them an equal amount of money, Lisa had thrice as much money as Joshua.
If Joshua had $42 in the end, how much money did Lisa have at first?

- A. $112
- B. $84
- C. $28
- D. $42

Question: 5 of 35

There were $3/7$ as many tables as chairs in a school hall. Mr. Chan added 60 tables and 60 chairs into the school hall for an event.
As a result, the number of tables was $3/5$ the number of chairs.
How many chairs were in the hall in the end?

- A. 100
- B. 200
- C. 220
- D. 250

Question: 6 of 35

The bar chart shows the number of students who took a variety of languages.

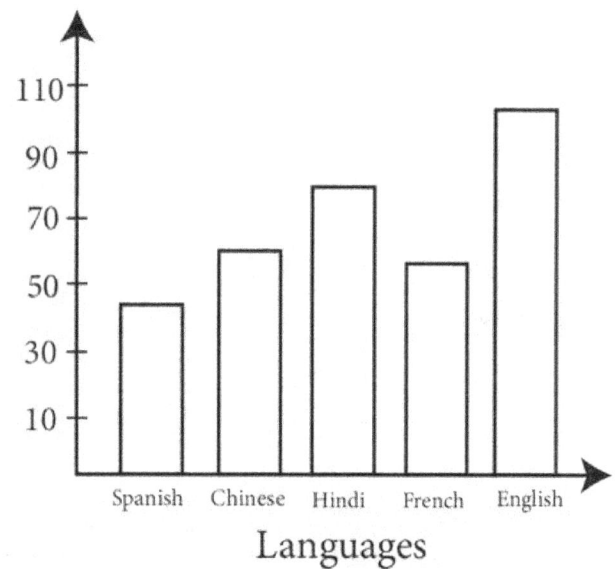

Which of the following is a correct statement?

- A. Students who took French and Chinese have the same amount.
- B. The difference between the students who took English and Hindi is 20
- C. The total of students who took English, Hindi, and, French is 640.
- D. The students who took Spanish are more than those who took French.

Question: 7 of 35

The line chart below shows the scores A in English in a variety of classes.

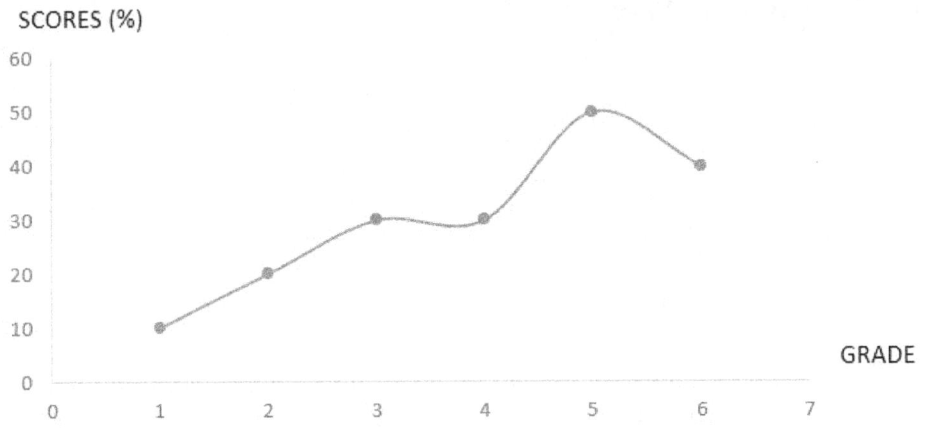

SCORES (%) vs GRADE:
- Grade 1: 10
- Grade 2: 20
- Grade 3: 30
- Grade 4: 30
- Grade 5: 50
- Grade 6: 40

Which of the following is a correct statement?

- A. The total scores of Grades 6 and 3 are 30.
- B. The total scores of Grades 1, 4, and 5 are 90
- C. The difference scores between Grades 2 and 1 is 45
- D. The difference scores between Grades 5 and 3 is 60.

Question: 8 of 35

Pinky Co makes pink paint by mixing red paint and white paint in a ratio of 3:4. Slacky Co makes pink paint by mixing red paint and white paint in a ratio of 5:7.
Which company uses a higher proportion of red paint in their mixture?

- A. They are the same
- B. Pinky Co
- C. Slacky Co
- D. It is impossible to tell

Question: 9 of 35

Below shows a bar chart.

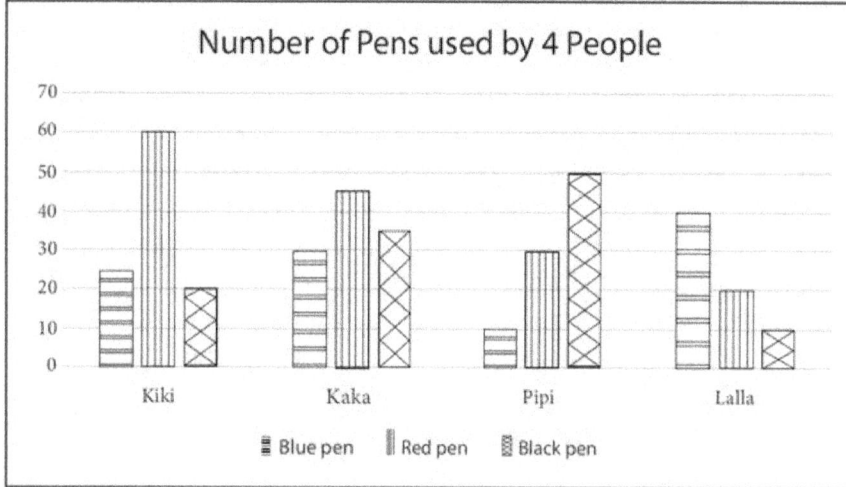

Which of the following is a correct statement?

- A. Kiki use more blue pen than Lalla
- B. Kaka uses more red pen than Kiki.
- C. Pipi uses more pens than Lalla.
- D. Kaka uses more black pens than Pipi.

Question: 10 of 35

The picture shows a graph.

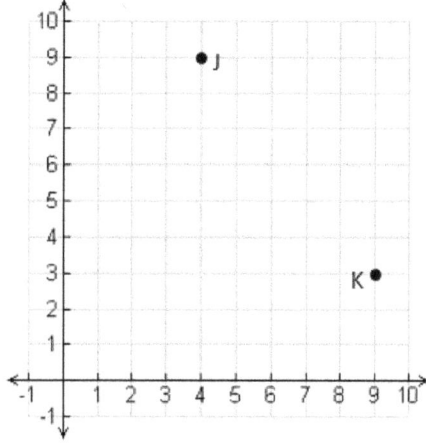

Points J, K, and M form a right-angled triangle. State the possible coordinates of M from the answer options.

- A. (4,3)
- B. (5,3)
- C. (6,9)
- D. (9,3)

Question: 11 of 35

A box $\frac{1}{2}$ filled with beans has a total mass of $8 \frac{2}{5}$ kg.

When the same box is completely filled with beans, the total mass is $15 \frac{8}{10}$ kg.
What is the mass of the box when it is empty?

- A. 1 kg
- B. 3.56 kg
- C. 3.9 kg
- D. 1.2 kg

Question: 12 of 35

A total of 32,280 people took part in the marathon. There were 5 times as many adults as children. When 30 women and 30 children withdrew from the marathon, the number of women who took part was twice the number of children who took part. Find the number of men who took part in the marathon.

- A. 55,256
- B. 1,190
- C. 16,550
- D. 16,170

Question: 13 of 35

The diagram below shows a line graph.

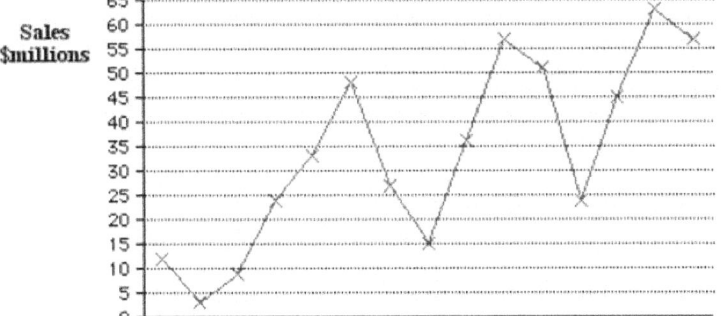

Which of the following is a correct statement?

- A. The sales in 1990 were lesser than the sales in 1993.
- B. The sales from 1991 to 1993 show an increasing pattern.
- C. The sales from 1997 to 1999 show an increasing pattern.
- D. The sales in 1995 were higher than the sales in 1999.

Question: 14 of 35

The diagram below shows a net of a cuboid.

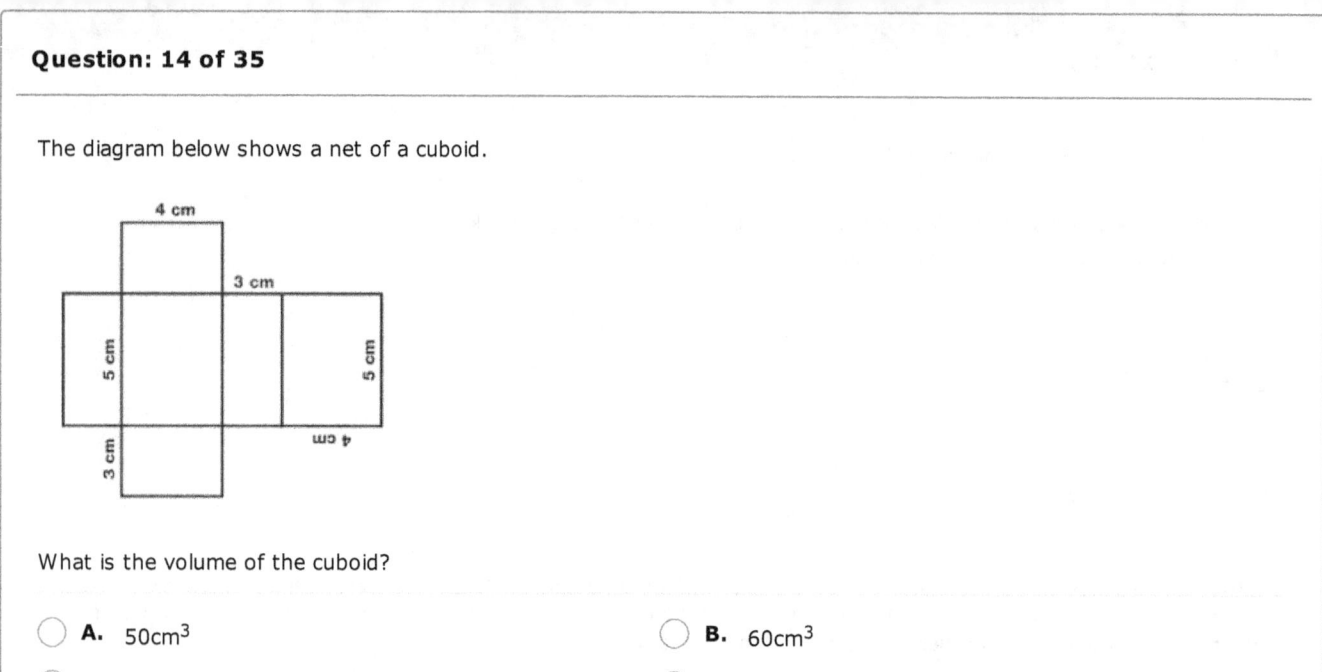

What is the volume of the cuboid?

A. 50cm³
B. 60cm³
C. 30cm³
D. 55cm³

Question: 15 of 35

The diagram below shows two rectangles.

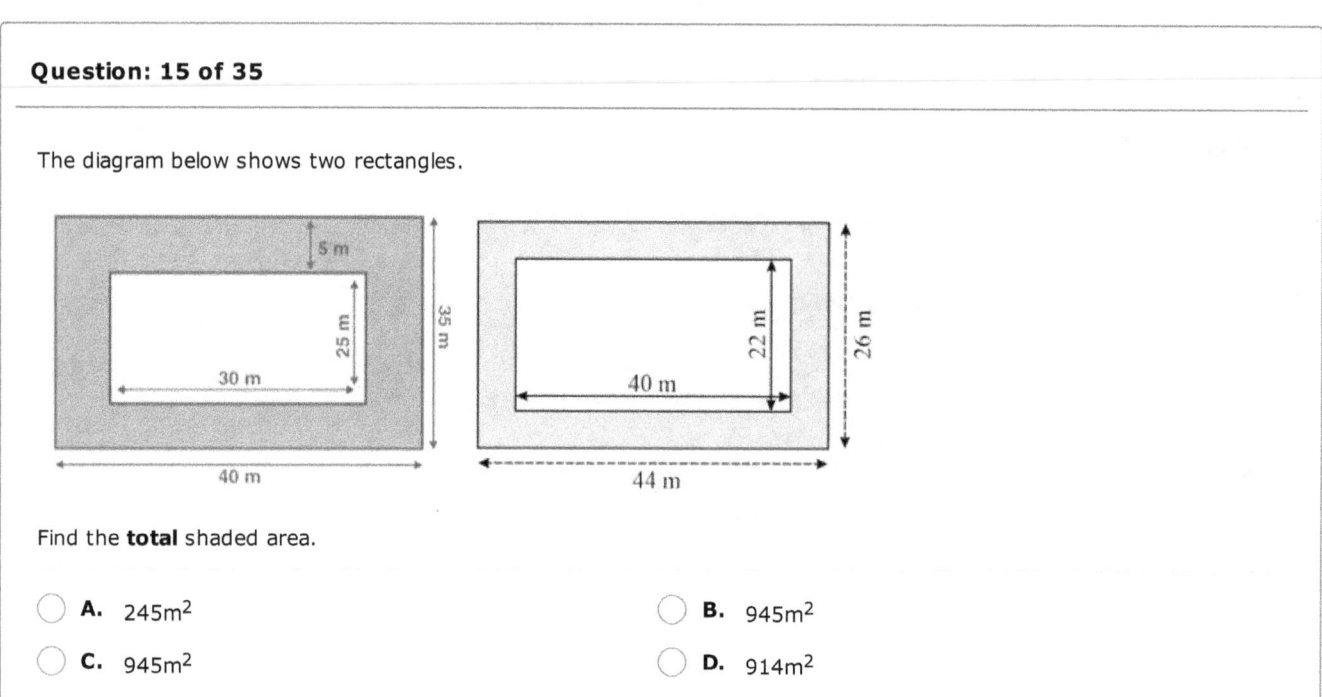

Find the **total** shaded area.

A. 245m²
B. 945m²
C. 945m²
D. 914m²

Question: 16 of 35

James prepared slices of mini cakes to be served during a party. When he gives 7 slices to each guest, he will have 2 slices left. When he gives 3 slices to each guest, he will have 134 slices left.
How many guests were at the party?

A. 33
B. 32
C. 43
D. 35

Question: 17 of 35

JK : Today, my age is 55 years old.
RM: 36 days later, my age will be the same as yours.
RJ : Oh, today is 6 March 2019!

According to the conversation above, what is RM's date of birth ?

- A. 10th April 1964
- B. 1st April 1954
- C. 14th April 1955
- D. 22nd April 1964

Question: 18 of 35

Shamila had some cookies. She sold $\frac{3}{4}$ of the cookies on the first day. She sold $\frac{1}{2}$ of the remaining cookies plus 2 more on the second day. She had 84 cookies left in the end.
How many cookies did Shamila have at first?

- A. 172
- B. 688
- C. 86
- D. 700

Question: 19 of 35

During a trivia quiz, points were awarded for questions answered as shown below.

Correct	5 Points
Wrong	-2 Points
Missed(Unanswered)	-1 Points

For Janice, the ratio of the questions answered correctly to questions answered wrongly to questions that were missed out was 9 : 2 : 1.

If Janice was awarded 360 points, how many questions are there in all?

- A. 80
- B. 100
- C. 110
- D. 108

Question: 20 of 35

Find the missing number based on the figure below.

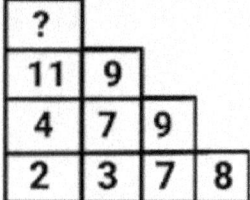

- A. 10
- B. 11
- C. 30
- D. 20

Question: 21 of 35

Amir started cycling from his house to town A at 9:00 a.m.

Along the way, he passed a pizzeria but didn't stop. He arrived in town A at 12:15 p.m. Calculate the average speed of the journey in km/h.

- A. 13.78 km/h
- B. 10.77 km/h
- C. 18.77 km/h
- D. 17.87 km/h

Question: 22 of 35

Jamie has some 50¢ coins and Kumar has some $1 coins. Jamie has 12 more coins than Kumar. The total amount of money Kumar has is $15 more than the total amount of money Jamie has. How many coins does Kumar have?

- A. 46
- B. 36
- C. 24
- D. 42

Question: 23 of 35

Some identical lightbulbs were hung on a rod 91 cm long. The first and the last lightbulb were hung 14 cm away from each end of the rod. The rest of the lightbulbs were hung at an equal distance of 9 cm apart.
How many lightbulbs were hung on the rod altogether?

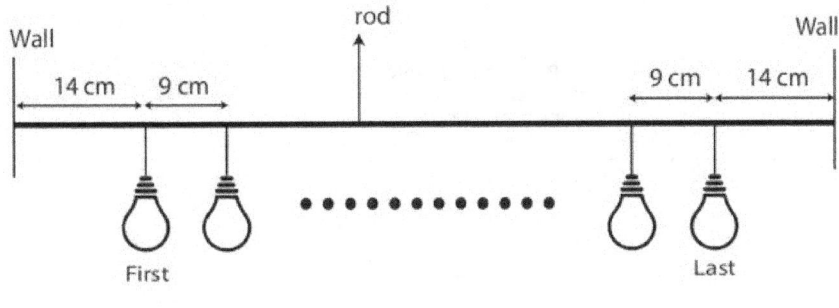

- A. 4
- B. 5
- C. 7
- D. 8

Question: 24 of 35

The graph shows the attendance of students to the school and average marks they got.

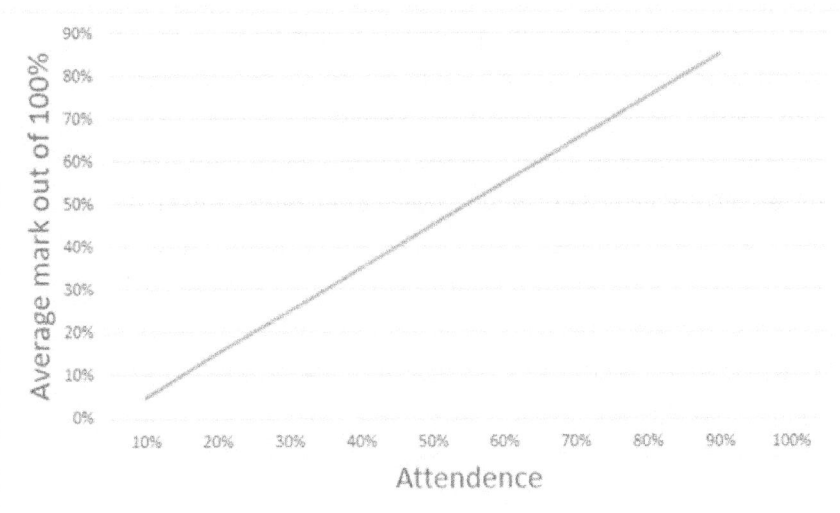

What is the average mark of the students who have an attendance of 40%?

- A. 65%
- B. 55%
- C. 15%
- D. 35%

Question: 25 of 35

Which of these statements is/are correct?

X. $2 - \frac{5}{9}$ is less than $1\frac{1}{3}$
Y. $\frac{1}{4} + \frac{3}{4}$ is equal to 1
Z. $\frac{5}{6}$ is less than $\frac{4}{3}$

- **A.** X and Y only
- **B.** Y and Z only
- **C.** Y only
- **D.** Z only

Question: 26 of 35

A resident of Los Angeles took a vacation to New Jersey. She leaves Los Angeles at 1400 on Tuesday for a 3-hour flight to New York. She spends 12 hours in New York before catching a 2-hour flight to New Jersey. If she spent two days in New Jersey, when did she leave for Los Angeles?

- **A.** 0500 Thursday
- **B.** 0500 Friday
- **C.** 0700 Thursday
- **D.** 0700 Friday

Question: 27 of 35

Every symbol represents a particular number and works as follows : each star represents 3; hence 3 x 3 =9 .

| ★ | ★ | 9 |

What's the largest number among the three numbers (**including one that needs to be calculated**) in the following case?

☀	☀	36
☁	☁	64
☀	☁	

- **A.** 100
- **B.** 64
- **C.** 36
- **D.** 128

Question: 28 of 35

Below is a set of 3 jugs. If jug Z has a capacity of 800 ml, how far will it be from being full (**remaining empty capacity**) after ½ of what's in jug X and $^1/_5$ of the content in jug Y is poured into it?

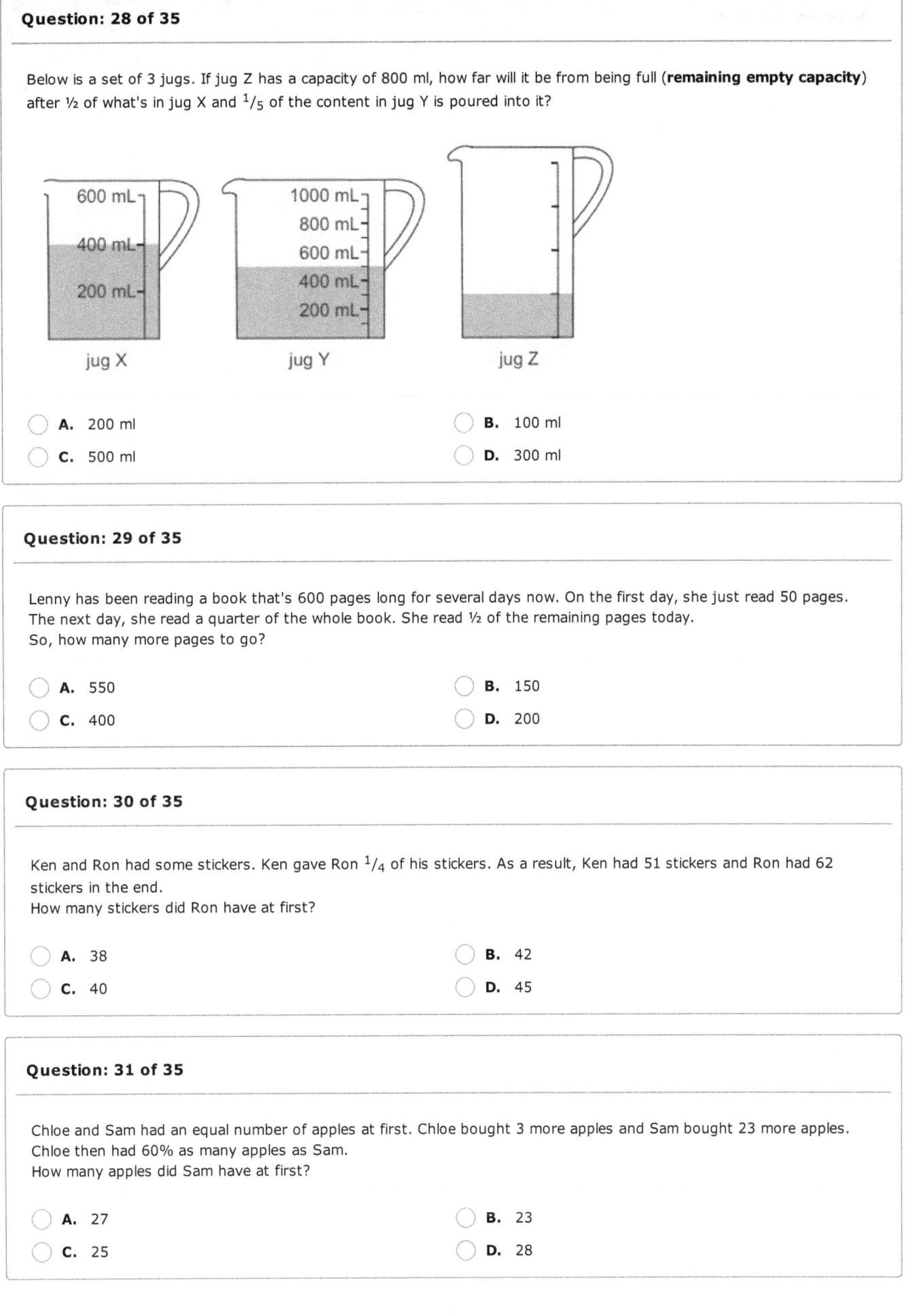

- A. 200 ml
- B. 100 ml
- C. 500 ml
- D. 300 ml

Question: 29 of 35

Lenny has been reading a book that's 600 pages long for several days now. On the first day, she just read 50 pages. The next day, she read a quarter of the whole book. She read ½ of the remaining pages today.
So, how many more pages to go?

- A. 550
- B. 150
- C. 400
- D. 200

Question: 30 of 35

Ken and Ron had some stickers. Ken gave Ron $^1/_4$ of his stickers. As a result, Ken had 51 stickers and Ron had 62 stickers in the end.
How many stickers did Ron have at first?

- A. 38
- B. 42
- C. 40
- D. 45

Question: 31 of 35

Chloe and Sam had an equal number of apples at first. Chloe bought 3 more apples and Sam bought 23 more apples. Chloe then had 60% as many apples as Sam.
How many apples did Sam have at first?

- A. 27
- B. 23
- C. 25
- D. 28

Question: 32 of 35

Teachers at Marybrown High School asked students to list their favorite desserts.

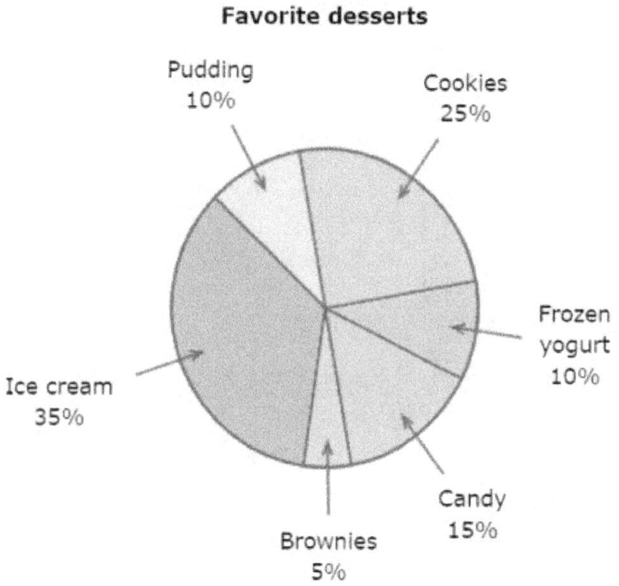

Favorite desserts
- Pudding 10%
- Cookies 25%
- Frozen yogurt 10%
- Candy 15%
- Brownies 5%
- Ice cream 35%

What is the measure of the central angle in the "Cookies" section?

- A. 45°
- B. 90°
- C. 120°
- D. 150°

Question: 33 of 35

A cold water and hot water tap supply water to the bathtub at the rate of 45 liters per minute and 15 liters per minute respectively.
The drain empties water at the rate of 12 liters per minute.
The volume of the tub is 624 liters.

When all the taps and drains are turned on at the same time (**tub was empty before**), how long will it take to overflow in minutes?

- A. 10 mins
- B. 12 mins
- C. 13 mins
- D. 15 mins

Question: 34 of 35

Karen had some muffins. She gave $1/6$ of them to Mary, 32 to Amy, and kept the remaining 13 muffins for herself. How many muffins did Karen have at first?

- A. 48
- B. 45
- C. 52
- D. 54

Question: 35 of 35

10 pupils were standing in a straight row at equal distances apart from each other. If the 2nd pupil was 20 m away from the 6th pupil, what was the distance between the first pupil and the last pupil?

- A. 40m
- B. 56m
- C. 45m
- D. 42m

Preptive Prepare Better
Series K Thinking Skill Trial Test

Name: _____

Date: _____

Question: 1 of 40

Detective Jackson investigates a theft at a museum. Three suspects, Amelia, Ben, and Chloe, provide clues:

Amelia: "I was at the opera with Ben."
Ben: "I was at the opera with Amelia."
Chloe: "I was at the library all evening."

Later, CCTV footage reveals only one culprit. It is not known if the culprit is one of the suspects. If both Amelia and Ben are telling the truth, who must be the thief?

- A. Amelia
- B. Ben
- C. Chloe
- D. Impossible to determine

Question: 2 of 40

A scientist studying a newly discovered element, "Element X," makes the following observations:

- No object is both magnetic and radioactive.
- All Element X samples are radioactive.

Based on these observations, which of the following is the most likely characteristic of Element X?

- A. Element X cannot be magnetic.
- B. Element X will always remain radioactive.
- C. Only Element X is radioactive.
- D. The radioactive property of Element X can be neutralized.

Question: 3 of 40

A bakery receives orders for three types of muffins: blueberry, cranberry, and chocolate chip. They process orders through a machine that follows these rules:

- Any order with "blueberry" in it outputs "breakfast muffin."
- Any order without "blueberry" but with "cranberry" or "chocolate chip" outputs "snack muffin."
- Any order without "blueberry," "cranberry," or "chocolate chip" outputs "plain muffin."

What would be the output for the following orders?

- Order 1: Blueberry and Cranberry
- Order 2: Chocolate Chip only
- Order 3: None of the known ingredients
- Order 4: Apple Fritter only

- A. Breakfast muffin, Snack muffin, Plain muffin, Plain muffin
- B. Snack muffin, Breakfast muffin, Plain muffin, Plain muffin
- C. Breakfast muffin, Breakfast muffin, Plain muffin, Plain muffin
- D. Snack muffin, Snack muffin, Plain muffin, Plain muffin

Question: 4 of 40

In the library, three books are displayed: "Mythology," "History," and "Science." You know the following:

- At least one of the books is fiction.
- "Mythology" is fiction if and only if "History" is non-fiction.
- "Science" is non-fiction if and only if "Mythology" is non-fiction.

Which of the following statements must be true?

- A. All three books are fiction.
- B. All three books are non-fiction.
- C. Only "Mythology" is fiction.
- D. None of the above.

Question: 5 of 40

A scientist conducts an experiment with three chemicals: X, Y, and Z. When mixed, they produce a specific color reaction:
- X and Y together produce no color.
- Y and Z together produce blue.
- X and Z together produce yellow.

Red and blue when mixed together create violet colour, and blue and yellow make green. What color will be produced when all three chemicals are mixed?

- A. Blue
- B. Yellow
- C. Green
- D. No color

Question: 6 of 40

A group of friends goes hiking. Each one takes same amout of time to hike. Sarah starts 30 minutes earlier than Emily, who starts 2 hours before John. If John takes 4 hours to complete the hike, how much earlier does Sarah finish compared to John?

- A. 1 hour 30 minutes
- B. 2 hours
- C. 2 hours 30 minutes
- D. 3 hours

Question: 7 of 40

A museum displays ancient artefacts with labels in three languages: English, French, and Spanish. If an artefact has a label in both English and French, only then it has a Spanish label. However, some artefacts only have labels in one or two languages.
Which of the following statements must be true?

- A. Every artefact with a Spanish label also has an English label.
- B. Every artefact with a French label also has a Spanish label.
- C. Every artefact with an English label also has a French label.
- D. Every artefact with at least one label has a Spanish label.

Question: 8 of 40

Sarah, Maya, and Daniel take turns watering three rose bushes, Rose A, Rose B, and Rose C. They follow a strict order: Sarah waters Rose A, then Maya waters Rose B, and finally, Daniel waters Rose C. After completing their rounds, they start again with Sarah watering Rose A.
If Sarah watered Rose A last Tuesday, who will water rose bush this coming Monday?

- A. Sarah
- B. Maya
- C. Daniel
- D. Impossible to determine

Question: 9 of 40

In a local library each book has a unique code starting with a capital letter (F for fiction, N for nonfiction, B for biography), followed by three digits representing its specific category within the broader classification.

If a book with code "F235" is about historical fiction, what can you think as possibility about the following book codes ?

- A. N142 – It can be a book about science.
- B. B589 – It can be a biography of a musician.
- C. F001 – It can be the first book in the fiction category.
- D. All of the above are possible.

Question: 10 of 40

In the realm of enchantments, there are three potions: Elixir of Wisdom, Mystic Tonic, and Shrouded Brew. The Potion Master leaves a note:
- Potion A: "Only the wise can discern the Elixir of Wisdom."
- Potion B: "Those who trust their inner-self and their sixth sense, may find solace in the Mystic Tonic."
- Potion C: "For those who navigate the shadows, the Shrouded Brew awaits."

If Alex is known for relying on intuition, which potion should Alex choose?

- A. Potion A
- B. Potion B
- C. Potion C
- D. Impossible to determine

Question: 11 of 40

In a distant galaxy, astronomers make a groundbreaking discovery of three celestial objects: Nebula X, Quasar Y, and Black Hole Z. Their observations reveal intriguing patterns:
- Pattern A: "Nebula X is always found near Quasar Y."
- Pattern B: "Black Hole Z exerts near gravitational influence over Nebula X."
- Pattern C: "Quasar Y emits energy pulses that Black Hole Z absorbs."

If astronomers locate Nebula X, which celestial object are they most likely to find nearby?

- A. Quasar Y
- B. Black Hole Z
- C. Impossible to determine
- D. Both Quasar Y and Black Hole Z

Question: 12 of 40

A group of time travelers embarks on a journey to three historical periods: Renaissance, Industrial Revolution, and Ancient Egypt. Their time machine follows specific guidelines:

- Guideline A: "The trip to Ancient Egypt precedes the visit to the Industrial Revolution."
- Guideline B: "If the Renaissance is the destination, the Industrial Revolution must follow."
- Guideline C: "The journey to Ancient Egypt is never the first destination."

Which historical period did the travellers visit first?

- A. Renaissance
- B. Industrial Revolution
- C. Ancient Egypt
- D. Impossible to determine

Question: 13 of 40

Intergalactic explorers—Stella, Victor, and Xavier—are on a quest for the legendary Galactic Treasure, navigating through cosmic challenges:
- Challenge A: "Stella often faces asteroid belt. If Stella encounters the asteroid belt, Victor must navigate through the gravitational anomaly immediately after."
- Challenge B: "Xavier, who is skilled in navigating gravitational anomalies, always precedes Stella in the treasure hunt."
- Challenge C: "If Victor successfully manoeuvres through the wormhole, Stella must encounter the asteroid belt immediately after."

If Xavier navigates through the wormhole, what possible challenge will Stella face next?

- A. Gravitational anomaly
- B. Asteroid belt
- C. Wormhole
- D. Impossible to determine

Question: 14 of 40

Observe the following pattern of the numbers.

[8, 6, 2, 7]
[9, 3, 6, 2]
[7, 8, Q, 5]
[5, 7, 4, 3]

Find the value of **Q**.

- A. 4
- B. 2
- C. 3
- D. 5

Question: 15 of 40

News Report: "There is a massive earthquake in parts of Indonesia."

Jerry: "My friend Colin is in danger. He lives in Indonesia."

Which of the following shows the mistake in Jerry's reasoning?

- A. There was a reported fatality in the Indonesian incident.
- B. Earthquake is a devastating happening.
- C. The Earthquake did not affect every part of the country
- D. The magnitude of this earthquake is high.

Question: 16 of 40

An examination's questions have the following mark allocation:
Q1→6 marks
Q2→4 marks
Q3→4 marks
Q4→4 marks
Q5→4 marks

If David got 12 marks for the paper, how many questions may he have answered correctly? (Students will be given either full or no mark for each question.)

- A. 4
- B. 2
- C. 3
- D. 1

Question: 17 of 40

Florence is sitting with 4 friends in a row on a bench. If there is an odd number of friends on either side, what is the position of Florence from the right?

- A. 2nd or 3rd
- B. 4th
- C. 2nd or 4th
- D. 3rd

Question: 18 of 40

Students are preparing for an excursion to the seaside. If Sam goes, Ray and Chris will not go, but Paul and Peter will go. If Pam goes, Stella and Basil go. Sam goes when Pam goes.

Pam is going to seashore. How many will go to the excursion among those mentioned?

- A. 5
- B. 4
- C. 3
- D. 6

Question: 19 of 40

Study the table of numbers below:

8	7	6	5	4
4	2	7	9	1
3	4	9	4	0

If column 3 and column 5 are interchanged and row 1 and row 3 are interchanged, what is the sum of the numbers in the 2nd column?

- A. 15
- B. 13
- C. 5
- D. 22

Question: 20 of 40

The following tables show the scores of four students in two Chemistry tests.

Test 1

Felix	71
George	72
Sally	69
Pam	53
Chris	61

Test 2

Felix	80
George	75
Sally	68
Pam	65
Chris	60

A prize is going to be given to the most improved student provided he/she has an average above 70 for both tests. Who should win the award?

- A. George
- B. Felix
- C. Sally
- D. Pam

Question: 21 of 40

John, the editor of a national newspaper needs to fill a vacancy for a Business Reporter in the newspaper. He needs someone who can write and investigate stories, is good with figures and economic terms and issues and also is readily available for frequent travels across the country.
He has shortlisted four applicants for this position:

i. Brandon: A married journalist, with five years of experience as a Business reporter. He has two young children and a wife, who is a nurse. Travel may be a problem for him.
ii. Ben, is a writer with about three years of experience as a travel blog writer. He is in his mid-20s and unmarried.
iii. Gray, has distinguished himself on the Business Desk of the local newspaper for the past three years, since he left the University four years ago. Loves to travel solo.
iv. Ray, who has a degree in Economics, has worked as a reporter for 4years on the Political Desk of another national newspaper.

Who should be offered the position?

- A. Brandon
- B. Ben
- C. Gray
- D. Ray

Question: 22 of 40

Dauphin is a new hairdresser in town. Since she opened her new shop three weeks ago, she found out patronage has not been encouraging as most women still consider her an outsider in town.

If you are consulting for her business what will you advise her to do to save her business?

- A. Start a new complimenting service to her hairdressing business.
- B. Join the local women's groups and get involved in community services.
- C. Start advertising her business on national newspaper.
- D. Relocating her business out of town.

Question: 23 of 40

The final of the football match will be played on the 7th Sunday after the opening match. The opening match was on Friday, June 23rd.
When will the final match be played?

- A. September 3
- B. August 13
- C. August 6
- D. July 30

Question: 24 of 40

If you go to bed late, you will wake up late next morning and miss your bus to work. If you miss your bus, you will get to work late. If you get to work late, $10 will be deducted from your wages that day.

If Paul gets to work late, which of the following is true?

- A. Paul left the house on time for his bus
- B. $10 was deducted from his day's wage
- C. Paul woke up early that day
- D. Paul slept early the night before.

Question: 25 of 40

A posthumous event is when it is held in celebration of someone's death. Which event below is the best example of a posthumous event?

- A. Musical legend, Sir Bob was awarded with a Lifetime Achievement Award last year.
- B. The great Egyptian Writer could not receive his third Breaker Award personally because he was in self-imposed exile.
- C. The Award for the deceased boxer was accepted on his behalf by his first son, who himself is a boxing champion.
- D. The Award ceremony was cancelled because the main winners boycotted the ceremony because of their belief against racism.

Question: 26 of 40

From the following statements, answer the following question.
I. Raymond is 8 years old.
II. Barney likes Mathematics
III. Barney is the same age as Raymond.
IV. Barney and Raymond are taught by the same teachers
V. Raymond likes the English Language

Which of the statements helps us to know that Raymond and Barney are possibly classmates?

- A. I, II, and III
- B. III and IV
- C. I, II, V
- D. All the above

Question: 27 of 40

The Grand City Museum hosts three prestigious annual exhibitions: Art, History, and Science. Each exhibition has a unique ticket price: Art ($25), History ($30), and Science ($20). A special "Museum Explorer Pass" grants access to all three exhibitions for a discounted price. The discount covers cost of at least one exhibition. However, the pass is only available if purchased online in advance.
Sarah wants to visit all three exhibitions and is debating between buying individual tickets or the online pass.
What is the maximum price Sarah will pay if she decides for the Museum Explorer Pass online?

- A. $50
- B. $55
- C. $60
- D. $65

Question: 28 of 40

A museum exhibit showcases historical artefacts from different civilizations. Each artefact has a label displaying its origin, date, and material.
If a label reads "Egyptian, 1450 BC, Gold," which of the following artefacts cannot be described by this label?

- A. A ceremonial mask
- B. A funerary statue
- C. A decorative glass vase
- D. A decorative pen

Question: 29 of 40

A group of friends, Emily, Michael, and Daniel, are planning a movie night. They each have different movie preferences: action, comedy, or drama. They share their thoughts:
- Emily: "I'm not in the mood for action tonight."
- Michael: "Drama isn't my thing, but I wouldn't mind a comedy."
- Daniel: "I'm open to action or comedy, but drama seems too slow."

Which movie genre will they most likely choose if they go together?

- A. Action
- B. Comedy
- C. Drama
- D. Any of the above

Question: 30 of 40

A city has three districts: East, West, and Central. Each district has a specific rule for crossing pedestrian bridges:
- East: You must be wearing a hat of a specific color.
- West: You must be carrying a specific type of flower.
- Central: You must be walking a dog.

You see Daniella crossing a bridge in the East district, not wearing a hat. Which of the following statements must be true?

- A. Daniella lives in the East district.
- B. Daniella is carrying a specific type of flower.
- C. Daniella is walking a dog.
- D. Daniella broke the rule for crossing the bridge in the East district.

Question: 31 of 40

Herminie and Ron send messages in a secret language. According to their language, "Be alert" is written as "Tr elaeb".

Ron sent "Tidnu of" to Herminie and what is the meaning of that?

- A. Found it
- B. Find it
- C. Found of
- D. None of the above

Question: 32 of 40

If there is no electricity, Adam lights candles.
If Adam has invited someone to dinner, he lights candles.
Adam doesn't use candles otherwise.

There wasn't a power failure, but Adam was lighting candles. That likely means..

- A. Adam has invited someone to dinner.
- B. Adam does not have invited someone to dinner.
- C. Cannot decide about Adam having dinner with someone.
- D. None of the above.

Question: 33 of 40

Four traffic lights at an intersection follow a specific pattern:
- Light A turns green and stays green for 30 seconds.
- Light B turns green 15 seconds after Light A and stays green for 30 seconds.
- Light C turns green 15 seconds after Light B and stays green for 30 seconds.
- Light D turns green 15 seconds after Light C and stays green for 30 seconds.

If you arrive at the intersection when Light A is just turning green, what is the minimum amount of time you will have to wait for all lights to be green at the same time?

- A. 2 minutes
- B. 1 minute 45 seconds
- C. 1 minute 30 seconds
- D. Not possible

Question: 34 of 40

Two friends, Maya and Ethan, each flip a coin. Maya claims her coin landed heads, while Ethan claims his coin landed tails. However, you know that one of them is always lying and the other is always telling the truth.
What can you definitively conclude?

- A. Maya's coin landed heads.
- B. Ethan's coin landed tails.
- C. At least one coin landed heads.
- D. None of the above.

Question: 35 of 40

A four-digit password is formed by using the digits 1, 2, 3, and 4. No digit can be repeated. You know that the first digit is odd, last digit is odd too, and the password is greater than 2000.
What could be the fourth digit?

- A. 1
- B. 2
- C. 3
- D. 4

Question: 36 of 40

Your body needs the vitamin B3 compound niacinamide to support the preservation of youthful-looking skin. Niacinamide, niacin and nicotinamide riboside are the three different forms of vitamin B3, however, niacinamide is the one that's most frequently seen in skincare products. Niacinamide makes your skin more moisturized and less sensitive by helping to lock in moisture and keep pollution or other potential irritants out. It has been demonstrated that niacinamide reduces inflammation, which can help lessen redness brought on by conditions including eczema, rosacea, and acne. Niacinamide can help keep your skin clear and smooth, which can reduce the appearance of pores and possibly cure dark spots, prevent skin cancer, and minimize fine lines and wrinkles.

If the information is true, which of the following must be false?

- A. Niacinamide helps in keeping your skin moisturized.
- B. Niacinamide can possibly treat dark spots.
- C. Niacinamide can be used by those with acne problems.
- D. Niacinamide is not ideal to be used topically on skin.

Question: 37 of 40

During summer time, all tourists at Lagoon beach are reminded to bring and drink water or any kind of refreshment while spending their time under the sun. The owner of the Lagoon beach says that it can help prevent the cases of fatality at Lagoon beach during summer time.

Which of the following situations best strengthens the reminder of the owner of the Lagoon beach?

- A. The owner of Lagoon beach wants to sell more of their refreshments especially during summer time.
- B. Cases of heat stroke that leads to death have happened at Lagoon beach during summer time over the past years.
- C. The safety officer of Lagoon beach wants to help the local vendors at Lagoon beach during summer time.
- D. The owner of Lagoon beach does not want any tourists at their beach during summer time.

Question: 38 of 40

Use the rule of the pattern and find the missing number which will replace the question mark.

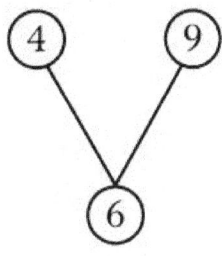 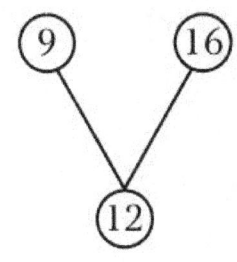 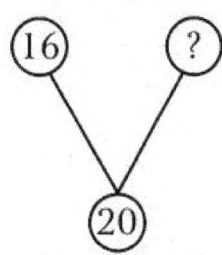

A. 21
B. 25
C. 45
D. 36

Question: 39 of 40

There are three separate large black boxes, and inside each large box there are two separate small red boxes; inside each of these small boxes, there is one smaller blue box.
How many boxes are there altogether?

A. 9
B. 12
C. 15
D. 18

Question: 40 of 40

Compare the knowledge of persons X, Y, Z, A, B and C in relation to each other. X knows more than A. Y knows as much as B. Z know less than C. A knows more than Y. The best knowledgeable person amongst all is:

A. C
B. X
C. A
D. Cannot be determined

Preptive Prepare Better
Series L Reading Comprehension Trial Test

Name: _____

Date: _____

Direction for Questions: 1 to 5

Read the two extracts and then answer the questions.

Extract A: Respect is Reciprocal

At the heart of Willowbrook stood an ancient oak tree, its sprawling branches reaching towards the heavens like outstretched arms. Beneath its shade, generations had gathered to seek solace, share stories, and reflect on the wisdom passed down through the ages. It was here that the tale of young Liam and his unexpected encounter unfolded, illustrating the profound truth of reciprocity in respect.

Liam, a spirited boy with a heart as boundless as the horizon, found himself at a crossroads one crisp autumn morning. As he wandered along the winding paths of Willowbrook, lost in thought, he stumbled upon a peculiar sight: an elderly gentleman struggling to carry a heavy load of firewood. Without hesitation, Liam rushed to his aid, offering a helping hand without expecting anything in return.

Moved by Liam's kindness, the elderly gentleman introduced himself as Mr. Thompson, a longtime resident of Willowbrook. Grateful for Liam's assistance, Mr. Thompson regaled him with tales of the town's history, imparting timeless lessons of respect, empathy, and reciprocity. As they parted ways, Mr. Thompson left Liam with a parting gift—a small acorn, symbolizing the seeds of wisdom planted through their encounter.

Inspired by Mr. Thompson's words, Liam embarked on a journey of self-discovery, determined to embody the principle of reciprocal respect in all his interactions. From offering a listening ear to a friend in need to volunteering at the local soup kitchen, Liam found joy and fulfillment in extending kindness to others, knowing that respect was a currency that enriched both giver and receiver.

As time passed, Liam's acts of kindness rippled throughout the community, fostering a culture of mutual respect and compassion. In return for his generosity, neighbors offered their support, friends stood by his side, and strangers became allies on life's winding path. The bond forged through reciprocity transcended mere gestures, weaving a tapestry of connections that bound the people of Willowbrook together as a unified whole.

In the end, it was beneath the ancient oak tree that Liam's journey came full circle. As he stood beneath its branches, surrounded by friends old and new, Liam reflected on the profound truth that had guided his steps: respect is indeed reciprocal. In giving of himself, Liam had received far more in return—a sense of belonging, purpose, and the enduring love of a community united by a shared belief in the power of respect.

Extract B: One Good Turn Deserves Another

In the bustling metropolis of Metro City, where the pace of life never slowed and opportunities beckoned from every corner, there existed a simple yet profound belief: "One good turn deserves another." Amidst the hustle and bustle of city life, the story of Emily and Marcus unfolded, illustrating the transformative power of kindness and reciprocity.

Emily, a young woman with a heart as vast as the city skyline, found herself in a predicament one rainy afternoon as she hurried through the crowded streets. Struggling to balance an armful of groceries while shielding herself from the downpour, Emily's umbrella gave way, leaving her drenched and disheartened. It was then that Marcus, a stranger passing by, offered his assistance without hesitation, sharing his umbrella and guiding her to shelter.

Grateful for Marcus's kindness, Emily thanked him profusely before parting ways. Little did she know, their chance encounter would set into motion a series of events that would forever change their lives. As fate would have it, Emily found herself in a position to pay it forward when she stumbled upon Marcus in need of assistance a few weeks later.

Marcus, a struggling artist trying to make ends meet, had encountered a setback in his pursuit of his passion. His studio space had fallen into disrepair, leaving him without a place to create and showcase his art. Recognizing an opportunity to reciprocate the kindness shown to her, Emily rallied her friends and neighbors to renovate Marcus's studio, transforming it into a vibrant space where creativity flourished.

Moved by Emily's gesture, Marcus found renewed inspiration in his art, creating masterpieces that captured the spirit of the city and touched the hearts of all who beheld them. In gratitude for Emily's generosity, Marcus used his talents to give back to the community, organizing art workshops for underprivileged youth and donating proceeds from his exhibitions to local charities.

As word spread of Emily and Marcus's story, the people of Metro City were inspired to embrace the principle of reciprocity in their own lives. Acts of kindness multiplied, fostering a culture of compassion, generosity, and goodwill that permeated every corner of the city. Strangers became friends, neighbors became allies, and the city became a beacon of hope and unity in a world often overshadowed by division and strife.

In the end, it was beneath the glow of the city lights that Emily and Marcus's journey came full circle. As they stood side by side, surrounded by the beauty of Marcus's artwork and the warmth of a community united in kindness, they realized the profound truth that had guided their paths: one good turn truly deserves another. In giving of themselves, Emily and Marcus had received far more in return—a sense of purpose, connection, and the enduring power of a city bound together by the transformative magic of reciprocity.

Question: 1 of 30

In Extract B, what phrase encapsulates the central belief in the transformative power of kindness?

- A. "Unity in diversity"
- B. "Every cloud has a silver lining"
- C. "One good turn deserves another"
- D. "Actions speak louder than words"

Question: 2 of 30

According to Extract A, what role does the ancient oak tree play in the community of Willowbrook?

- A. It serves as a gathering place for social events
- B. It provides shade for weary travelers
- C. It symbolizes the passage of time and wisdom
- D. It represents the town's commitment to environmental conservation

Question: 3 of 30

How does Liam respond to Mr. Thompson's tales in Extract A?

- A. He becomes bored and walks away
- B. He listens attentively and learns valuable lessons
- C. He interrupts Mr. Thompson repeatedly
- D. He falls asleep while Mr. Thompson speaks

Question: 4 of 30

What change occurs in Marcus's life as a result of Emily's gesture in Extract B?

- A. He decides to leave Metro City permanently
- B. He becomes more selfish and withdrawn
- C. He gains renewed inspiration in his art
- D. He loses interest in pursuing his passion

Question: 5 of 30

What central message about reciprocity is conveyed through both extracts?

- A. Kindness should only be extended to those who deserve it
- B. Acts of generosity often lead to negative consequences
- C. Reciprocity enriches both the giver and the receiver
- D. It is important to expect something in return for acts of kindness

---

Direction for Questions: 6 to 11

Read the two extracts and then answer the questions.

Extract A: The Greatest Artwork: A Journey Through Time and Perception

Art, in its myriad forms, has captivated and inspired humanity for centuries, transcending cultural boundaries and epochs. The quest to identify the greatest artwork is a subjective endeavor, influenced by personal taste, historical context, and the elusive essence of artistic genius. As we embark on this journey through time and perception, we encounter masterpieces that have left an indelible mark on the canvas of human history.

One cannot discuss the greatest artwork without acknowledging Leonardo da Vinci's enigmatic masterpiece, the Mona Lisa. Painted in the early 16th century, this iconic portrait continues to intrigue and beguile viewers with its enigmatic smile and subtle nuances. Da Vinci's meticulous attention to detail, masterful use of sfumato technique, and psychological depth imbue the Mona Lisa with a timeless allure that transcends its physical dimensions.

Moving beyond the realms of painting, the greatest artwork also encompasses monumental achievements in sculpture, such as Michelangelo's David. Carved from a single block of marble in the early 16th century, this towering statue exemplifies the pinnacle of Renaissance artistry. Michelangelo's ability to breathe life into stone, capturing the essence of human anatomy and emotion, elevates David to a status of unparalleled grandeur and significance.

In the realm of architecture, the Taj Mahal stands as a testament to the power of love and the zenith of Mughal craftsmanship. Commissioned in the 17th century by Emperor Shah Jahan as a mausoleum for his beloved wife, Mumtaz Mahal, this architectural marvel mesmerizes visitors with its symmetrical beauty, intricate marble inlay work, and ethereal aura. The Taj Mahal transcends its function as a tomb, embodying a timeless tribute to enduring love and artistic excellence.

Beyond the confines of traditional mediums, the realm of literature offers its own contenders for the title of the greatest artwork. Shakespeare's Hamlet, often hailed as the pinnacle of English literature, is a timeless exploration of existential angst, moral ambiguity, and the complexities of the human condition. Through its richly layered characters, poetic language, and profound philosophical insights, Hamlet continues to resonate with audiences across centuries and continents.

As we navigate the vast landscape of artistic expression, the question arises: what defines the greatest artwork? Is it the technical virtuosity of the artist, the emotional resonance of the piece, or its enduring impact on society? Perhaps it is a combination of all these factors, intertwined with the ineffable qualities that elevate certain works of art to the realm of greatness.

Ultimately, the greatest artwork transcends the constraints of time and space, speaking to something fundamental within the human experience. It challenges us to contemplate the nature of beauty, truth, and the fleeting passage of time. Whether captured on canvas, chiseled in stone, or penned on parchment, the greatest artwork serves as a beacon of inspiration, inviting us to glimpse the sublime and ponder the mysteries of existence.

Extract B: The Most Valuable Art: Exploring Worth Beyond Price Tags

In a world where the value of art is often measured in monetary terms, it is imperative to recognize that true worth transcends mere price tags. While the art market may assign staggering values to certain works based on factors such as rarity, provenance, and aesthetic appeal, the most valuable art is often found in the intangible realms of emotion, cultural significance, and human connection.

One cannot discuss the most valuable art without acknowledging the transformative power of works that challenge societal norms and inspire social change. From Picasso's Guernica, a searing indictment of war and violence, to Banksy's politically charged murals, art has the ability to provoke thought, spark dialogue, and galvanize movements for justice and equality. The true worth of these works lies not in their market value, but in their capacity to effect meaningful change in the world.

Similarly, the most valuable art encompasses works that offer solace, comfort, and hope in times of adversity. Whether it's the stirring melodies of Beethoven's Symphony No. 9 or the timeless wisdom of literature such as Viktor Frankl's Man's Search for Meaning, art has the power to uplift the human spirit and provide solace in the face of adversity. The value of these works lies not in their price at auction, but in their ability to offer solace, inspire resilience, and remind us of the enduring strength of the human spirit.

Moreover, the most valuable art extends beyond the confines of traditional mediums and prestigious galleries, encompassing the rich tapestry of cultural heritage and folk traditions from around the world. From indigenous crafts passed down through generations to vibrant street art that adorns city walls, art in its myriad forms serves as a celebration of diversity, heritage, and collective memory. The true worth of these works lies not in their marketability, but in their ability to preserve cultural identity, foster community, and enrich the fabric of society.

In essence, the most valuable art defies quantification, transcending the limitations of commerce and embracing the boundless realms of human creativity and expression. It is found not only in the masterpieces that adorn museum walls but in the everyday acts of creativity, resilience, and compassion that shape our shared human experience. As we navigate the complexities of the art world, let us remember that true worth cannot be measured in monetary terms alone, but in the profound impact that art has on hearts, minds, and souls.

Question: 6 of 30

What distinguishes the Mona Lisa and David as discussed in Extract A?

- A. Their simplicity and minimalist design
- B. The intricate details and symbolism infused by the artists
- C. Their significant monetary value in the art market
- D. The use of vibrant colors and bold brushstrokes

Question: 7 of 30

What aspect of Shakespeare's Hamlet is highlighted in Extract A?

- A. Its role as a timeless exploration of existential themes
- B. Its popularity as the most widely performed play
- C. Its portrayal of traditional values and customs
- D. Its use of innovative theatrical techniques

Question: 8 of 30

According to Extract A, what defines the greatest artwork?

- A. Its monetary value in the global art market
- B. The technical proficiency of the artist
- C. Its emotional resonance and enduring impact
- D. The popularity and fame of the artist

Question: 9 of 30

How does Extract B emphasize the role of art in times of adversity?

- A. By discussing its monetary value as an investment
- B. By analyzing its influence on fashion and design trends
- C. By highlighting its capacity to offer solace and hope
- D. By comparing it to traditional cultural artefacts

Question: 10 of 30

How does Extract B portray the value of indigenous crafts and street art?

- A. As commodities with high market demand
- B. As reflections of cultural identity and heritage
- C. As symbols of elitism and exclusivity
- D. As products of mainstream art institutions

Question: 11 of 30

According to Extract B, how does art contribute to societal dialogue and activism?

- A. By conforming to established cultural norms and values
- B. By serving as a status symbol for the elite class
- C. By encouraging passive consumption and enjoyment
- D. By provoking thought, sparking dialogue, and galvanizing movements for justice

Direction for Questions: 12 to 17

Read the extracts and then answer the questions.

Extract A:
Narrative writing, a captivating form of literary expression, transcends the mere conveyance of stories; it's an artistry that breathes life into characters and orchestrates plots as intricate as a ballet performance. Each stroke of vivid description and every line of compelling dialogue coalesce to create a symphony that resonates in the readers' hearts and minds. It's a voyage through uncharted territories, whether navigating a spine-tingling mystery, relishing a heartwarming romance, or traversing the fantastical landscapes of epic sagas. Narrative writing beckons readers to transcend reality and immerse themselves in the enchanting realm of storytelling, where imagination reigns supreme and emotions run wild.

Extract B:
Expository writing, a beacon of informative discourse, stands as a testament to clarity and enlightenment. It transcends the confines of mere explanation, embodying a scholarly pursuit to elucidate complex concepts and dissect intricate ideas with precision. Each sentence, meticulously crafted, serves as a gateway to a world of knowledge and understanding. Like an erudite mentor, expository writing navigates through the labyrinth of information, shedding light on multifaceted subjects and empowering readers with insights that unravel the mysteries of the world.

Extract C:
Persuasive writing, an influential force commanding attention, wields the tools of reason, emotion, and eloquence to sway opinions and ignite action. It's a skillful manipulation of language and rhetoric, orchestrating a symphony of words to captivate minds and stir souls. Embedded within speeches, advertisements, or impassioned editorials, persuasive writing assumes the role of a catalyst, igniting ideological shifts and propelling readers towards decisive convictions and actions. It's a dynamic dance between rhetoric and conviction, seeking not just agreement but resonance, leaving an indelible mark on the fabric of beliefs.

Extract D:
Creative writing, an expansive universe brimming with diversity across genres, serves as a testament to boundless imagination and artistic exploration. It transcends the conventions of traditional storytelling, becoming a canvas for linguistic experimentation and emotional exploration. It births verses that encapsulate fleeting moments with poignant precision and narratives that challenge the norms, inviting readers into realms unknown. Creative writing, an ode to the limitless possibilities of language, invites both creators and audiences on a transcendental journey where innovation meets expression, birthing worlds anew with each stroke of the pen.

Question: 12 of 30

Which extract emphasizes the artistry of storytelling, comparing it to a ballet performance?

- A. Extract A
- B. Extract B
- C. Extract C
- D. Extract D

Question: 13 of 30

In which extract is writing portrayed as a canvas for linguistic experimentation and emotional exploration?

- A. Extract A
- B. Extract B
- C. Extract C
- D. Extract D

Question: 14 of 30

Which extract likens narrative writing to a symphony that resonates in readers' hearts and minds?

- A. Extract A
- B. Extract B
- C. Extract C
- D. Extract D

Question: 15 of 30

In which extract is writing described as an influential force commanding attention and igniting action?

- A. Extract A
- B. Extract B
- C. Extract C
- D. Extract D

Question: 16 of 30

Which extract portrays writing as a pursuit to elucidate complex concepts and empower readers with insights?

- A. Extract A
- B. Extract B
- C. Extract C
- D. Extract D

Question: 17 of 30

Which extract highlights a form of writing that focuses on unravelling complex concepts and providing clear insights to its readers?

- A. Extract A
- B. Extract B
- C. Extract C
- D. Extract D

Direction for Questions: 18 to 24

Seven sentences have been removed from the text. Choose from the sentences (A – H) the one which fits each gap (1 – 7). There is one extra sentence which you do not need to use.

Eco-tourism

1.... The insatiable demand for travel experiences often comes at the expense of our planet, leading to pollution, habitat destruction, and cultural exploitation. Ecotourism, however, offers a promising alternative, promoting responsible travel that benefits both the environment and local communities.

At its core, ecotourism is defined as "responsible travel to natural areas that conserves the environment and improves the well-being of local people."2....., They are: minimizing environmental impact, respecting local cultures, supporting local communities, promoting environmental education etc.
3...... By supporting ecotourism initiatives, travelers can contribute to the conservation of biodiversity and endangered species, protection of natural resources, empowerment of local communities, promotion of cultural understanding and appreciation. Ecotourism fosters intercultural exchange and helps to break down barriers between different cultures, increasing environmental awareness and action.

Despite its significant advantages, ecotourism is not without challenges. Ensuring that ecotourism practices are truly sustainable requires careful planning, implementation, and monitoring. 4.......

As travelers become increasingly conscious of their environmental impact, ecotourism is poised to play a crucial role in shaping the future of the tourism industry. 5.....
As we move forward, it is essential to recognize that ecotourism is not simply a niche tourism segment but a fundamental shift towards responsible travel. 6.....The journey towards a more sustainable future begins with each individual traveler's choice, and ecotourism offers a powerful pathway for positive change.

7......Because it is going hand in hand with the sustainable development.

- A. **This definition encompasses a set of principles that guide ecotourism practices**
- B. **The benefits of ecotourism are multifaceted and far-reaching.**
- C. **By choosing ecotourism experiences, travelers can actively contribute to a more. sustainable and equitable world, ensuring that future generations can continue to enjoy the beauty and wonder of our planet.**

D. By embracing ecotourism principles and supporting initiatives that uphold these values, we can create a tourism industry that is both economically viable and environmentally and socially responsible.
E. Issues such as overtourism, greenwashing, and inadequate community involvement. need to be addressed to ensure the long-term success of ecotourism initiatives.
F. Ecotourism has many disadvantages.
G. In an age of increasing environmental awareness, the traditional tourism model is at a critical juncture.
H. Ultimately, ecotourism is very important for the future.

Question: 18 of 30

Which one fits in (1) ?

A. Sentence A
B. Sentence B
C. Sentence G
D. Sentence H

Question: 19 of 30

Which one fits in (2) ?

A. Sentence A
B. Sentence B
C. Sentence C
D. Sentence D

Question: 20 of 30

Which one fits in (3) ?

A. Sentence A
B. Sentence D
C. Sentence B
D. Sentence E

Question: 21 of 30

Which one fits in (4) ?

A. Sentence A
B. Sentence B
C. Sentence F
D. Sentence E

Question: 22 of 30

Which one fits in (5) ?

A. Sentence F
B. Sentence B
C. Sentence C
D. Sentence D

Question: 23 of 30

Which one fits in (6) ?

A. Sentence A
B. Sentence B
C. Sentence C
D. Sentence D

Question: 24 of 30

Which one fits in (7) ?

- **A.** Sentence A
- **B.** Sentence G
- **C.** Sentence H
- **D.** Sentence C

Direction for Questions: 25 to 27

Read the poem then answer the questions.

The Listeners
by **Walter De La Mare**

'Is there anybody there?' said the Traveller,
Knocking on the moonlit door;
And his horse in the silence champed the grasses Of the forest's ferny floor:
And a bird flew up out of the turret,
Above the Traveller's head
And he smote upon the door again a second time;
'Is there anybody there?' he said.
But no one descended to the Traveller;
No head from the leaf-fringed sill
Leaned over and looked into his grey eyes,
Where he stood perplexed and still.
But only a host of phantom listeners
That dwelt in the lone house then
Stood listening in the quiet of the moonlight
To that voice from the world of men:
Stood thronging the faint moonbeams on the dark stair,
That goes down to the empty hall,
Hearkening in an air stirred and shaken
By the lonely Traveller's call.
And he felt in his heart their strangeness,
Their stillness answering his cry,
While his horse moved, cropping the dark turf,
'Neath the starred and leafy sky;
For he suddenly smote on the door, even
Louder, and lifted his head:—
'Tell them I came, and no one answered,
That I kept my word,' he said.
Never the least stir made the listeners,
Though every word he spake
Fell echoing through the shadowiness of the still house
From the one man left awake:
Ay, they heard his foot upon the stirrup,
And the sound of iron on stone,
And how the silence surged softly backward,
When the plunging hoofs were gone.

Question: 25 of 30

What is the tone of the poem "The Listeners"?

- **A.** Joyful
- **B.** Mysterious
- **C.** Angry
- **D.** Sad

Question: 26 of 30

What does the poet convey through the line "That I kept my word"?

- A. The Traveller kept his promise.
- B. The Traveller broke his promise.
- C. The Traveller was lying.
- D. The Traveller was angry.

Question: 27 of 30

What is the deeper meaning of the poem "The Listeners"?

- A. The importance of communication
- B. The significance of keeping promises
- C. The mystery of existence and unanswered questions
- D. The loneliness of the Traveller

Direction for Questions: 28 to 30

Read the poem and then answer the questions.

Speak Kindly
by **Kate McKinney**

Speak kindly in the morning,
When you are leaving home,
And give the day a lighter heart
Into the week to roam.
Leave kind words as mementoes
To be handled and caressed,
And watch the noon-time hour arrive
In gold and tinsel dressed.
Speak kindly in the evening!
When on the walk is heard
A tired footstep that you know,
Speak one refreshing word,
And see the glad light springing
From the heart into the eye,
As sometimes from behind a cloud
A star leaps to the sky.
Speak kindly to the children
That crowd around your chair,
The tender lips that lean on yours
Kiss, smooth the flaxen hair;
Someday a room that's lonesome
The little ones may own,
And home be empty as the nest
From which the birds have flown.

Question: 28 of 30

What does the poet mean by "leave kind words as mementoes"?

- A. Kind words are important
- B. Kind words are essential
- C. Kind words can be remembered by others
- D. None of the above

Question: 29 of 30

What is the significance of the phrase "watch the noon-time hour arrive in gold and tinsel dressed"?

- A. Kind words can turn ordinary things into something beautiful
- B. Kind words are useless
- C. Kind words are essential
- D. None of the above

Question: 30 of 30

What does the poet mean by "smooth the flaxen hair"?

- A. Running hand through a child's hair
- B. Running hands through a boy's hair
- C. Running hands through an adult's hair
- D. None of the above

Preptive Prepare Better
Series L Mathematical Reasoning Trial Test

Name: _____

Date: _____

Question: 1 of 35

The **average** mass of Alex, Ben and Charles is 49 kg.
Alex is 9 kg heavier than Ben and 6 kg heavier than Charles.

What is Charles' mass?

- A. 44 kg
- B. 45 kg
- C. 48 kg
- D. 50 kg

Question: 2 of 35

Pablo can choose ham, turkey, tomato, or cheese on his sandwich.
If he chooses two different toppings, how many different sandwiches can he make, assuming order of toppings matters?

- A. 6
- B. 12
- C. 24
- D. 8

Question: 3 of 35

There were an equal number of boys and girls in the Art club at first. Then, another 17 boys joined the club and another 7 girls left the club.
There are four times as many boys as girls in the club now.
How many girls were there at first?

- A. 15
- B. 25
- C. 24
- D. 28

Question: 4 of 35

Mrs Lim bought 4 kg of strawberries.
Li Ming ate $\frac{1}{4}$ of it and Ming Hui ate $\frac{3}{5}$ kg.
How many kilograms of strawberries were left?

- A. $2\,^3/_5$ kg
- B. $2\,^2/_5$ kg
- C. $2\,^1/_2$ kg
- D. $1\,^2/_5$ kg

Question: 5 of 35

Ema had 4 times as many beads as Ben at first.
After Ema gave 60 beads to Ben, both of them had the same number of beads.

How many beads did each of them have at first?

- A. 180, 30
- B. 160, 20
- C. 150, 40
- D. 160, 40

Question: 6 of 35

A bakery shop baked 10.05 kg of cookies. Some of the cookies were packed into small packets of 250 g each and the rest was packed into big packets of 600 g each. In the end, the number of big packets was 4 more than the number of small packets.
How many big packets of cookies were packed?

- A. 13
- B. 18
- C. 23
- D. 33

Question: 7 of 35

The table shows the length of four ribbons.

Ribbon	A	B	C	D
Length (cm)	11.3	6.5	?	?

The average length of the 4 ribbons is 9.6 cm.
Write down one possible set of lengths for Ribbon C and Ribbon D.

- A. 10 cm, 11.6 cm
- B. 11 cm, 9.6 cm
- C. 8.2 cm, 12.9 cm
- D. None of these

Question: 8 of 35

In July, Raees, Leon and James saved a total of $1200. In August, Raees doubled his savings, Leon decreased his savings by $160 and James increased his savings by $110. Their savings were the same in August.
What was Raees' savings in August?

- A. $360
- B. $220
- C. $720
- D. $460

Question: 9 of 35

ABCD is a trapezium. ACD is an isosceles triangle where AD = DC. ∠ABC = 74° and ∠ACB = 86°. Find ∠ADC.

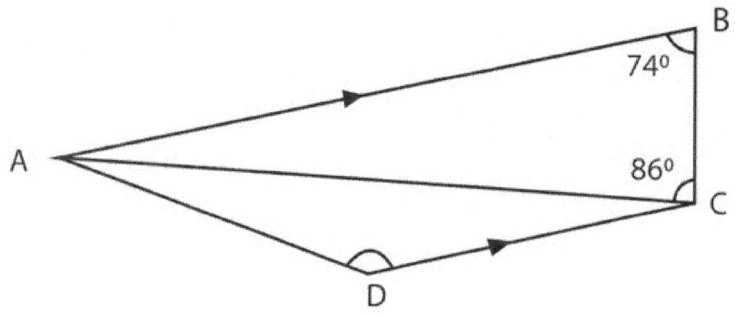

- A. 140°
- B. 100°
- C. 120°
- D. 160°

Question: 10 of 35

The diagram below shows a bar chart.

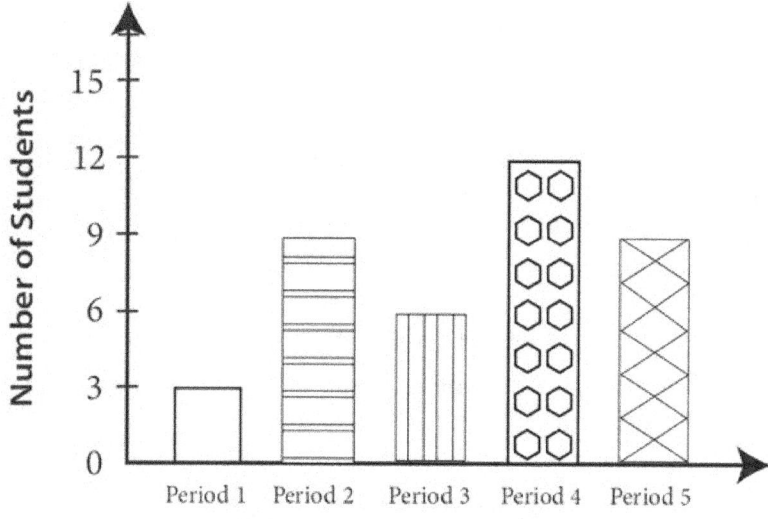

Which of the following is a false statement?

- A. The number of students in all of the class categories is 39
- B. The difference in the number of students between Period 4 and period 5 is 3
- C. The total number of students in periods 3, 4 and 1 is 21
- D. The difference in the number of students between the highest and the lowest is 10

Question: 11 of 35

Kelly bought some stickers. He gave $^4/_6$ of them to his brother and $^2/_5$ of the remainder to his cousin. After that, Kelly had 66 stickers left. How many stickers did Kelly buy?

- A. 330
- B. 210
- C. 120
- D. 190

Question: 12 of 35

The figure below shows a rectangle ABCD. The ratio of AB to BC is 3 : 2. AB is 90 cm. What is the perimeter of rectangle ABCD?

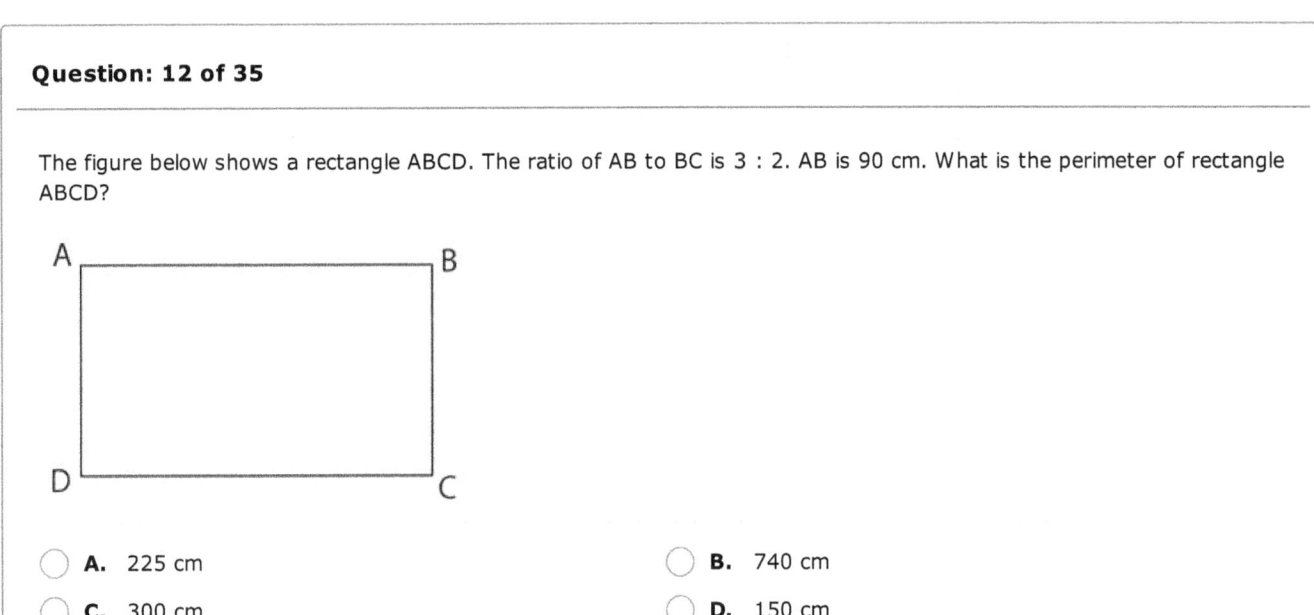

- A. 225 cm
- B. 740 cm
- C. 300 cm
- D. 150 cm

Question: 13 of 35

The pie chart shows the results of a survey conducted to identify the favorite game of some students.

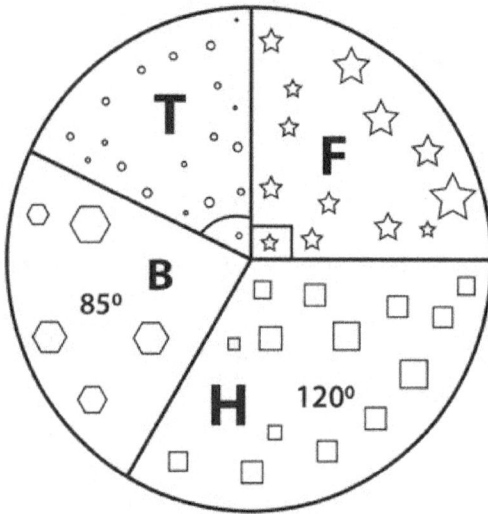

F Football   H Hockey   B Badminton   T Tennis

How many students like Tennis if the total number of students is 360?

A. 100

B. 90

C. 55

D. 65

Question: 14 of 35

The given figure shows a party cap.

What is the cross-section obtained when a horizontal cut parallel to the base is given to the cap?

A. Circle

B. Cone

C. Triangle

D. Rectangle

Question: 15 of 35

Yoga classes were held for the following duration in a particular week.

Day	Duration of exercise (min)
Sunday	40
Monday	50
Tuesday	60
Wednesday	x
Thursday	30
Friday	90
Saturday	90

If the average duration of yoga class in the week is 60 minutes, what was the duration of the yoga class on Wednesday?

- A. 60 min
- B. 30 min
- C. 55 min
- D. 65 min

Question: 16 of 35

Four friends - Dan, Fran, Ian, and Jan - went strawberry picking last Saturday.

Strawberries picked by four friends

represents 5 punnets of strawberries

Dan
Fran
Ian
Jan

The pictograph shows the number of punnets of strawberries each of them picked. If this information is displayed instead as a pie chart, what would be the angle (to the nearest degree) of the sector of the pie chart representing the strawberries picked by Jan?

- A. 25°
- B. 50°
- C. 65°
- D. 82°

Question: 17 of 35

What is the y-intercept of the following line?

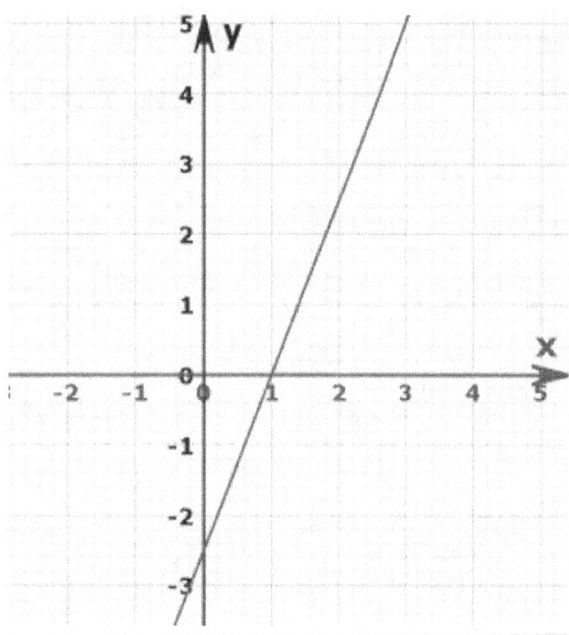

- A. (0, -2.5)
- B. (0, - 1)
- C. (0, 1)
- D. (0, 2.5)

Question: 18 of 35

Four classes 5A, 5B, 5C and 5D donated money at a Charity event.
5A, 5B and 5C donated a total of $380. 5C and 5D donated a total of $208.
The ratio of the amount donated by 5C to the total amount donated by the four classes is 1 : 6.
What is the amount donated by 5C?

- A. $44
- B. $64
- C. $72
- D. $84

Question: 19 of 35

Wesley and Xavier have 217 marbles altogether. Xavier and Yixian have 105 marbles altogether. Wesley has 3 times as many marbles as Yixian.
How many marbles does Xavier have?

- A. 59
- B. 112
- C. 56
- D. 49

Question: 20 of 35

The working hours are 9:00 am to 5:00 pm, and in between 60 minutes are spent on lunch. Find the ratio of time spent on work to the time spent on lunch.

- A. 8:30
- B. 7:1
- C. 7:5
- D. 12:7

Question: 21 of 35

Ms Nami runs the school cafeteria. On April, she had to order **15 cases of caramel pudding twice** due to the high demanding from her students. **Each contains p** of caramel pudding.

Choose the expressions that represent how many caramel puddings Ms Nami ordered on April.

- A. 30p
- B. 2p + 15p
- C. p(2 + 15)
- D. 3(15p)

Question: 22 of 35

A candy dispenser put various numbers of orange candies into bags.

Orange candies per bag

Stem	Leaf
2	4
3	0 3 5
4	4 6
5	5 6 6 7
6	4 5 6
7	3 4 7 9
8	2 5 5 5 6 8 8
9	0

How many bags had at least 20 orange candies?

- A. 30
- B. 29
- C. 27
- D. 25

Question: 23 of 35

The table shows the number of beads in a container.

Color	Number of Beads
White	180
Red	160
Blue	$\frac{2}{5}$ of the number of white beads

Daisy uses $1/4$ white beads and 20% of red beads to embroider her dress.

Count the number of beads left in the container.

A. 275
B. 298
C. 335
D. 333

Question: 24 of 35

Which of the following statements are correct?

X. $2/3 + 2/3$ is more than 1 ½
Y. $1 - 2/6$ is less than $1/6$
Z. $1/5$ is more than $1/12$

A. x and y only
B. y and z only
C. y only
D. z only

Question: 25 of 35

Emmy is redecorating her house. She decides to paint the half of the wall white. On top of the white paint, she paints orange, extending up to $1/5$ of the unpainted wall. As for the rest of the wall, she settles for yellow.

If the wall was divided into 10 equal parts, what would be the difference between the part painted orange only and yellow?

A. 1 part
B. 4 parts
C. 3 parts
D. 5 parts

Question: 26 of 35

These are measure of time spent a day in various rooms by Joe. Data is for 2 days. A day has 24 hrs.

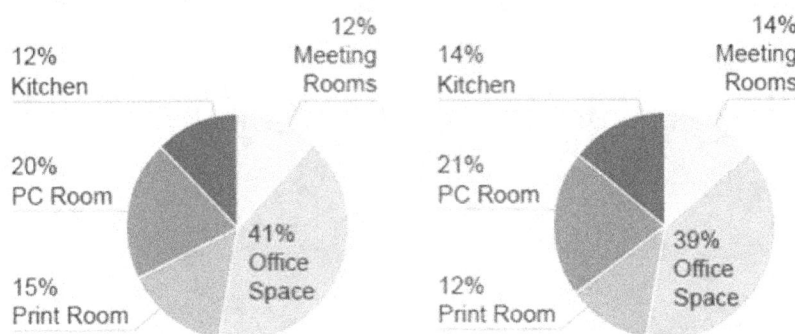

What's the difference between time spent in the kitchen between the first and second days, in hours ?

A. 0.48 hrs
B. 0.64 hrs
C. 0.12 hrs
D. 0.24 hrs

Question: 27 of 35

Kevin baked some pies.
He sold $\frac{1}{3}$ of the pies in the morning. He sold $\frac{1}{4}$ of the remainder in the afternoon.
He had 156 pies left. How many pies did Kevin bake?

A. 302
B. 212
C. 312
D. 324

Question: 28 of 35

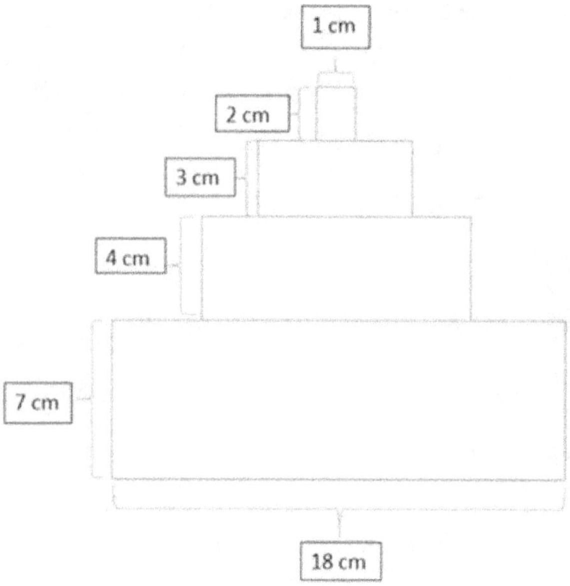
1 cm
2 cm
3 cm
4 cm
7 cm
18 cm

Find the perimeter of the above figure.

A. 68 cm

B. 70 cm

C. 72 cm

D. 74 cm

Question: 29 of 35

Below shows the prices of some items at a bookshop.

888
1 for $25.80

4 for $5.40

Kenny bought 2 calculators and 16 notebooks for $60.30. There was a **discount given on the calculators only**. What was the percentage discount of the calculators?

A. 25%

B. 50%

C. 75%

D. 40%

Question: 30 of 35

The average of three 2-digit numbers is 45. One of the numbers is 70.
What is the largest difference between the other two numbers?

A. 55

B. 45

C. 65

D. 25

Question: 31 of 35

In the figure below, find the sum of ∠a, ∠b, ∠c, ∠d, ∠e, and ∠f.

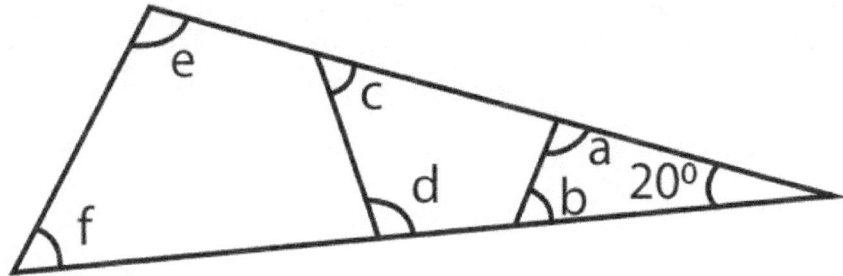

A. 160°

B. 240°

C. 360°

D. 480°

Question: 32 of 35

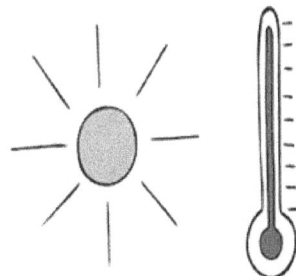

The average temperature of a town for the first 4 days of a month was 58°. The average temperature of the second, third, fourth and fifth day of this week was 60°.
If the ratio of the temperature of the first day and fifth day of the week is 7:8, then what was the temperature of the fifth day?

A. 62°

B. 52°

C. 64°

D. 72°

Question: 33 of 35

Alex and Belle had $88. Alex saved $1/3$ of his money while Belle spent $4/5$ of her money. Given that Alex and Belle spent the same amount of money, how much did Alex save?

- A. $16
- B. $12
- C. $14
- D. $18

Question: 34 of 35

Mei placed some potted plants in a row from one end to the other end of the corridor. They were placed at an equal distance from one another. The distance between the 9th and the 13th potted plant was 24 m. If the length of the entire stretch of corridor was 78 m, find the total number of plants that were placed altogether.

- A. 14m
- B. 16m
- C. 15m
- D. 12m

Question: 35 of 35

Nurul and Mei Shan had an equal number of sweets at first. After Nurul gave away 192 sweets and Mei Shan gave away 24 sweets, Mei Shan had 4 times as many sweets as Nurul.

How many sweets did each girl have at first?

- A. 244
- B. 225
- C. 248
- D. 256

Preptive Prepare Better
Series L Thinking Skill Trial Test

Name: _____

Date: _____

Question: 1 of 40

Paul's mum: If you pass the Mathematics test, you can go on an excursion to the seaside.

Paul didn't go on an excursion.

Which of the following is the most correct conclusion?

- A. Paul didn't pass the Mathematics test
- B. Paul passed the Mathematics test
- C. Paul passed the Mathematics test but decided not to go for the excursion
- D. Can't decide

Question: 2 of 40

In a video game, each game won gives the players the same number of points which gradually leads to a payout level. There are 4 payout levels in the game.

Level 1: The number of games won is less than that of Level 2.
Level 2: The number of games won is less than Level 3
Level 3: The number of games won is less than Level 4
Level 4: The number of games won is more than Level 3

Based on the information, can you determine ...

- A. Number of games won in Level 2 is double those in 1
- B. The number of games won in Level 3 is double the same in Level 1
- C. Inadequate information to determine the number of games in each level
- D. The number of games won in Level 4 is double those of Level 2

Question: 3 of 40

Jack Carlos has just been appointed the interim coach of Tradford United. He will be in charge for the rest of the season for the next 12 matches.
If he manages to qualify the club for Intercontinental competition by the end of the season, he will be confirmed as the chief coach for next season. Otherwise, a new coach will be appointed for next season.

Pedro: "I came to know a new coach won't be appointed for Tradford United next season. That means Carlos did a good job as the interim coach."
Ben: "Carlos can only serve as the interim coach for the rest of the season until a new coach (not Carlos) is appointed for the team next season, no matter what is team's result."

Which of the following reasoning is true?

- A. Pedro Only
- B. Ben Only
- C. Both Pedro and Ben
- D. Neither Pedro nor Ben

Question: 4 of 40

The football competition's final is billed as the most spectacular sporting event of the decade.
There was a massive advertisement in the media with promotional billboards everywhere, and the gate fees were discounted. A full stadium is expected.
But just two hours before the gate was opened a heavy rainstorm started. The roads were flooded and the stadium became waterlogged. Authorities advised everyone to stay at home, and residents of the city agreed.

Barry: "The fans' turnout will be disappointing with this weather."
George: "There will still be a large fan turnout with the extensive promotion for the match."

Which of the following reasoning is most likely correct?

- A. Barry Only
- B. George Only
- C. Both Barry and George
- D. Neither Barry nor George

Question: 5 of 40

Some friends were trying to recall the number of the bus they took from the city centre to the seaside three months ago.
They can remember it was a B Bus as it travelled westward out of town to the seaside. The three friends agreed that the bus was an afternoon bus with a 4 and the second to the last number was 7.

If the bus number comprises whether its route is within the city (A) or outside (B). morning (1-), afternoon (2-) or evening (3-) ; and three other numbers, which is the most probable number of the bus the friends took to the seaside?

- A. B1-047
- B. A2-477
- C. A1-741
- D. B2-470

Question: 6 of 40

A thief steals a painting from a museum, leaving a cryptic note: "The moon reveals the truth, before darkness sets in." Police discover two hidden doors in the museum, one facing East and one facing West.
Given that:
- The full moon rises in the east.
- Security cameras are activated at sunset.

Through which door did the thief escape?

- A. East door, before cameras activated.
- B. East door, after cameras activated.
- C. West door, before cameras activated.
- D. West door, after cameras activated.

Question: 7 of 40

A bakery offers muffins in three flavors: blueberry, cranberry, and chocolate chip. Orders include the flavor and additional topping options: nuts, sprinkles, or none. The following are customer requests:
- Order 1: Blueberry with nuts
- Order 2: Cranberry, no topping
- Order 3: Chocolate chip, sprinkles

The bakery accidentally mixes up the toppings. Which order received the wrong topping?

- A. Order 1
- B. Order 2
- C. Order 3
- D. Impossible to determine

Question: 8 of 40

A city has three bus routes: Red, Green, and Blue; each covers specific areas: North, West, and East, respectively. Given the following information:
- The Red route stops at the City Hall in the North.
- The Green route serves the Museum in the East.
- The Blue route goes past the Train Station in the West.

If a passenger boards a bus at the Library, located in the West, which route should they take to reach close to the Airport in the East?

- A. Red
- B. Green
- C. Blue
- D. Possible with any route

Question: 9 of 40

A hotel offers three room types: Standard, Deluxe, and Suite. Each has a different view: city, park, or ocean. Given the following information:
- Standard rooms never have an ocean view.
- Deluxe rooms offer either a city or park view.
- Suites always have an ocean view.

If a guest specifically requests a room with a park view, which type of room should they book?

- A. Standard
- B. Deluxe
- C. Suite
- D. Impossible to guarantee

Question: 10 of 40

A library organizes books by genre (fiction, non-fiction) and publication date (old, new). Each section has specific borrowing limits: 2 books for fiction (old or new), 3 books for non-fiction (old or new).
A student borrows one old fiction book and two new non-fiction books. Did they break any borrowing rules?

- A. Yes, exceeding the fiction limit.
- B. Yes, exceeding the non-fiction limit.
- C. No, within all borrowing limits.
- D. Information insufficient to determine.

Question: 11 of 40

Archaeologists discover a stone tablet with an inscription in an unknown language. They identify recurring symbols and decipher their meanings:
- A circle represents "sun"
- A triangle represents "water"
- A square represents "land"
- A line represents "path".

The inscription reads: "Circle, triangle, line, square, triangle, square, circle." What does the inscription likely convey?

- A. A journey from land to water by a sunlit path.
- B. A cyclical relationship between sun, water, and land.
- C. A warning about dangers related to sun, water, and land.
- D. Instructions for a ritual involving sun, water, and land.

Question: 12 of 40

An explorer finds a faded treasure map with cryptic symbols:
- A skull – "Danger lies ahead."
- A compass pointing north – "Follow the true north."
- A palm tree – "Seek the shade of the tallest."
- A chest – "Your reward awaits beyond the fallen giant."

The explorer finds a skull-shaped rock marking a trail leading north. After following it, they reach a clearing with three palm trees, one significantly taller than the others. Nearby, a fallen tree trunk lies partially buried. Where should the explorer dig?

- A. Beneath the skull-shaped rock, assuming it marks the starting point.
- B. Under the tallest palm tree, following the "shade" clue literally.
- C. At the base of the fallen tree trunk, interpreting "fallen giant" metaphorically.
- D. More information is needed about the surrounding area and terrain.

Question: 13 of 40

Ginseng has been utilized to enhance general health. Additionally, it has been used to boost immunity, aid in illness prevention, and reduce stress. Ginseng of many varieties has been used to treat diabetes, erectile dysfunction in men, and unclear thinking. Ginseng may lower your risk of contracting the flu or a cold. Ginseng users frequently report feeling more alert. Ginseng may enhance performance in tasks requiring mental computation, focus, memory, and other abilities.

If the information is true, which of the following is false?

- A. Ginseng enhances and sharpens your memory.
- B. Ginseng enhances your brain's function.
- C. Ginseng makes you more alert making you restless.
- D. Ginseng prevents illness and boosts immunity.

Question: 14 of 40

Which of the following completes the series of figures.

- A.
- B.
- C.
- D.

Question: 15 of 40

If you start a journey facing north, walk for 6 km, turn right, walk for 2 km, turn left, and walk for 10 km, which direction will you be facing eventually?

- A. West
- B. East
- C. South
- D. North

Question: 16 of 40

If you wake up late, you won't have breakfast. If you miss breakfast, you will faint during the school assembly. If you faint, you will miss the morning classes. If you miss the morning classes, you won't see the new teacher.

If Angela missed her morning classes, which of the following is **false** ?

- A. Angela woke up early but missed her breakfast
- B. Angela took her breakfast
- C. Angela saw the new teacher
- D. Angela didn't faint during the assembly

Question: 17 of 40

A sports team is trying to qualify for the playoffs. They have played 30 games so far and won 12 games, and they need to win at least 60% of their games to secure a spot.

If the information in the bold letters is true, whose reasoning is correct?

Lisa: "If we win 18 of the remaining 20 games, we will secure a spot in the playoffs."
Jake: "To secure a spot in the playoffs, we need to have won at least 48 out of the total 50 games played."

- A. Lisa only
- B. Jake only
- C. Both Lisa and Jake
- D. Neither Lisa nor Jake

Question: 18 of 40

Mark argues that using solar energy is the most effective solution to combat climate change.
"Solar energy is renewable and produces no greenhouse gas emissions," Mark asserts.
Which one of these statements, if true, most weakens Mark's argument?

- A. The production and disposal of solar panels can have negative environmental impacts, contributing to pollution.
- B. Solar energy is not consistently available in all geographic regions, making it unreliable as the sole energy source.
- C. Fossil fuel energy sources are becoming more efficient and cleaner, reducing their environmental impact.
- D. Wind energy has been found to be more cost-effective and efficient than solar energy in generating electricity.

Question: 19 of 40

John argues that implementing a four-day workweek for employees is the best solution to improve work-life balance and employee productivity. "A shorter workweek will allow employees more time for personal activities and reduce burnout," John asserts.

Which one of these statements, if true, **most weakens** John's argument?

- A. Employees may feel more stressed to complete the same amount of work in a shorter period, leading to increased burnout.
- B. A shorter workweek has been linked to higher job satisfaction levels among employees.
- C. Implementing a four-day workweek may result in reduced pay for employees, leading to financial strain.
- D. Other companies that have adopted a four-day workweek have reported increased employee productivity and satisfaction.

Question: 20 of 40

If the product is of high quality, it must have passed rigorous testing. It is mandatory requirement.

If this is true, which one of these sentences must also be true?

- A. If the product has passed rigorous testing, it must be of high quality.
- B. If the product hasn't passed rigorous testing, it must not be of high quality.
- C. If the product has passed rigorous testing, it cannot be of high quality.
- D. If the product hasn't passed rigorous testing, it must be of high quality.

Question: 21 of 40

During the Christmas period a store offers a special promotion on its household products on sale with discounted prices for the products. This was based on a code attached to the price tag. Shoppers are expected to add the two numbers attached to the price tag and subtract their sum from the original price to find the selling promo price of the items.
For example, a $20 item has "23", attached, implying that the discount percentage is $5/20 = 25\%$

From the following 3 items, which one offers the most percentage of discounts to shoppers?
- Item A: $32, "35"
- Item B: $20, "42"
- Item C: $48, "75"

- A. Item A
- B. Item B
- C. Item C
- D. All offer the same discounts

Question: 22 of 40

Tessy is looking for a suitable location to open her new gift store specializing in children's birthday gifts. She is looking for a shop not too far from the main street in town and one that would attract the right kind of foot traffic.

Which of the following locations shown to her by the realtor is appropriate for her needs?

I. A store in a small mall located just on the outskirts of town, with two other gift stores and a saloon in the mall.
II. A big store on the ground floor of a new high-rise building about 20 minutes from the centre of town.
III. A little shop about two blocks away from the town's main street, located across the street from a popular small grocery mall and beside a female saloon.
IV. A stand-alone big store on a quiet residential street 20 minutes away from the town's centre

- A. I Only
- B. II Only
- C. III Only
- D. IV Only

Question: 23 of 40

Your cousin just sent you a message. He is sending a gift to your father for his 60th birthday. It is a special gift.
"It has a face and two hands but no arms or legs. But its hands cannot clap."

What can be the gift?

- A. A shirt
- B. A painting
- C. A clock
- D. A smartphone

Question: 24 of 40

The lock to a box is opened by a set of 4-alphabetic letters given by the following clues:
- the first letter starts the car
- The second letter ends the whole chant
- The third starts the whole chant
- The fourth starts the race
- And the whole is the emperor.

Which set of letters is the code that opens the lock?

- A. K, E, Y, S
- B. C, Z, A, R
- C. B, E, S, T
- D. D, A, R, T

Question: 25 of 40

A newspaper publisher assigns unique identification numbers to its newspaper using the following system:
- The first two digits represent the year it is published
- The third and fourth represent the quarter
- The fifth and sixth represent month
- The last two digits represent the day of the month published

If a newspaper's ID number is 82020523, which of the following statements must be false?

- A. It may be published in 1982.
- B. It was published in June
- C. It was published in the second quarter
- D. It was published on the 23rd of a month

Question: 26 of 40

There is a serial burglar on the loose on the block. The Policemen investigating the series of break-ins got the description of the suspect from 4 residents, who claimed to have seen the suspect leaving some of the apartments that were broken into.
-Witness 1: The suspect is a Latino male, in his mid-twenties, 5'3", with short dark hair and a broad chest. He was wearing dark blue jeans and a blue T-shirt.
-Witness 2: The suspect is a Mexican Male, aged 45-50 years old, 5'10", 245 pounds, with a shaved head. He was wearing a navy blue blazer and brown trousers.
-Witness 3: The suspect is dark-skinned, male, approximately 25 years old, 5'9", slight build, with short dark hair. He was wearing a blue suit.

During a Police surveillance of the block, a suspect was caught breaking into an apartment.

Description of suspect: He is a dark-skinned male, aged 24 years old, 140 pounds, about 5'9", with short dark hair. He was wearing a dark blazer and a pair of brown trousers.

Based on the descriptions by the three earlier witnesses, which of these witness descriptions was most likely that of the caught suspect?

- A. Witness 1
- B. Witness 2
- C. Witness 3
- D. Any of them

Question: 27 of 40

A restaurant offers three types of burgers: classic, bacon cheeseburger, and veggie burger. The menu states:
• All burgers come with fries.
• Bacon cheeseburgers do not include a drink.
• Veggie burgers come with either a salad or onion rings.

Which option is impossible?

- A. Ordering a classic burger with a salad.
- B. Ordering a bacon cheeseburger and a drink.
- C. Ordering a veggie burger with fries and onion rings.
- D. Ordering a classic burger without fries.

Question: 28 of 40

Three trains, the Comet, the Express, and the Local, travel between City A and City B. Their travel times are unique:
- The Comet takes 4 hours.
- The Express takes 5 hours.
- The Local takes 8 hours.

If two trains leave City A simultaneously heading for City B and arrive at their destination with a 3-hour difference, which trains were they?

- A. Comet and Express
- B. Comet and Local
- C. Express and Local
- D. Information insufficient

Question: 29 of 40

Four friends, Alice, Bob, Charlie, and Damien, live in a building with four floors. Each friend lives on a different floor, and nobody shares a floor. They gave clues about their residences:
- Alice lives neither on the top nor the bottom floor.
- Bob doesn't live on the ground floor.
- Charlie is two floors above Damien.

Who lives on which floor?

- A. Alice – 2nd, Bob – 4th, Charlie – 3rd, Damien – 1st
- B. Alice – 3rd, Bob – 2nd, Charlie – 1st, Damien – 4th
- C. Alice – 2nd, Bob – 1st, Charlie – 4th, Damien – 3rd
- D. Alice – 3rd, Bob – 4th, Charlie – 2nd, Damien – 1st

Question: 30 of 40

A group of friends organized a potluck party. Each person agreed to bring a dish that starts with a specific letter of the alphabet, ensuring a diverse spread. Here are their assignments:
- Ann: Appetizing salad
- Boy: Bread for lunch
- Claire: Casserole
- Donny: Desserts like apple pie
- Emma: Entrée like lasagna

However, on the day of the party, some changes occurred:
Ann brought a salad.
Boy brought sweet dessert instead of bread.
Claire couldn't find a casserole recipe and brought chips.
Donny baked an apple pie.
Emma, true to her assignment, brought a delicious lasagna.

Which dish does not fulfil either the alphabetical assignment or the food type assignment?

- A. Salad
- B. Sweet dessert
- C. Chips and dip
- D. Apple pie

Question: 31 of 40

A health scientist tells that:

"Even though chemical fertilizers and pesticides are some of the main developments in the agricultural sector, they are a major cause of water pollution, and they seriously affect our kidneys too."

Which of these statements, if true, best supports the scientist's claim?

- **A.** Organic fertilizers and natural pesticides are way better and more effective than chemical fertilizers and pesticides.
- **B.** When people drink that polluted water, the chemicals in chemical fertilizers and pesticides impair their kidneys.
- **C.** When walking on agricultural fields, chemical fertilizers and pesticides may go through the wounds on the skin, which may cause kidney failures.
- **D.** Chemical fertilizers and pesticides can be recycled.

Question: 32 of 40

If James improves his coding skills, he's likely to become a more efficient programmer. Becoming a more efficient programmer will help him complete projects faster. The only other way James can complete projects faster is by attending coding workshops.

Which one of the following is not possible?

- **A.** James improved his coding skills and became a more efficient programmer, which helped him complete projects faster.
- **B.** James did not improve his coding skills, but he attended coding workshops and started completing projects faster.
- **C.** James did not improve his coding skills, did not attend coding workshops, and could not complete projects faster.
- **D.** All are possible

Question: 33 of 40

Mike: "To become proficient in a foreign language, you should practice speaking with native speakers."
Lisa: "I had a conversation with a native speaker last week, so I'm fluent now."

Which sentence best points out the incorrect **assumption behind** Lisa's argument?

- **A.** Regular conversations with native speakers enhance language skills.
- **B.** Lisa's fluency in the language has nothing to do with speaking with native speakers.
- **C.** Speaking occasionally with native speakers guarantees language proficiency.
- **D.** Mike's advice about practising with native speakers is inaccurate.

Question: 34 of 40

"To excel in basketball, one needs to have height, agility and good coordination."

Mark: "Sarah is tall and has excellent coordination. She's destined to excel in basketball."
Lily: "However, Sarah lacks the agility and quick reflexes that are also crucial for basketball success. She can't excel in basketball"

Whose reasoning is correct?

- A. Mark only
- B. Lily only
- C. Both Mark and Lily
- D. Neither Mark nor Lily

Question: 35 of 40

If a laptop is a gaming laptop, it is expensive.
George bought an expensive laptop yesterday.

What is the conclusion about George's laptop ?

- A. It is a gaming laptop
- B. It isn't a gaming laptop
- C. It is expensive but not a gaming laptop
- D. No conclusion

Question: 36 of 40

There are three special cards in a deck: "Star" "Moon," and "Sun."
The following rules apply:
- The "Star" is superior if and only if "Moon" is not on display.
- "Sun" is superior even if "Moon" is displayed.
- "Moon" is superior only if the "Sun" is hidden.

Which of the following statements must be true?

- A. The least valuable card is "Moon"
- B. All three cards are superior.
- C. Only "Star" is a valuable card.
- D. "Sun" is the most valuable card.

Question: 37 of 40

In a display of various polygons and numbers, the reveals were in the following order:
- hexagon (12)
- pentagon (10)
- rectangle (8)

What will be the logical next shape and accompanying figure?

- A. Octagon (14)
- B. Square (4)
- C. Triangle (6)
- D. Heptagon (3)

Question: 38 of 40

There are special cards that can be drawn in a game:
- The Blue card gives you an addition of 5 marks;
- The Red card deducts 5 marks from your total;
- The Yellow card can alternate with an addition or subtraction of 5 marks each time you draw it, with a subtraction first.

Based on these rules, what will be your marks addition or subtraction if you draw a Yellow card, a Red card, a Red card, and a Blue card in that order?

- A. +10
- B. -10
- C. +15
- D. -15

Question: 39 of 40

Four friends, Mia, Noah, Olivia, and Peter, are discussing their favorite movie genres:
- Mia loves comedies and animation.
- Noah is a big fan of dramatic action and thrilling sci-fi movies.
- Olivia prefers dramas and documentaries on current affairs.
- Peter enjoys action-packed thrillers and mysteries.

There's a new movie festival with four categories: Action, Comedy, Documentary, and Thriller. Each friend can only choose one movie to watch.
Knowing their preferences, which movie category is least likely to be chosen by any of them?

- A. Action
- B. Comedy
- C. Documentary
- D. Thriller

Question: 40 of 40

Identify the wrong term.
14, 41, 52, 25, 36, 63, 74, 47, 56, 85

- A. 74
- B. 47
- C. 56
- D. 85

Preptive Prepare Better
Series I Reading Comprehension Trial Test

Answers and Solutions

Q: 1   What is the central theme of Extract A?

- A. The importance of authenticity in content creation
- ✓ Correct Ans **B. The role of storytelling in digital content**
- C. The significance of collaboration among content creators
- D. The impact of data-driven decision-making in content strategy

Explanation

Extract A primarily discusses the significance of storytelling as a central aspect of content creation in the digital landscape.

Q: 2   According to Extract A, what essential role does adaptability play in content creation?

- A. It limits the scope of content diversity
- B. It ensures consistent adherence to trends
- ✓ Correct Ans **C. It enables creators to navigate evolving platforms and algorithms**
- D. It hampers collaboration opportunities with other creators

Explanation

Extract A highlights the necessity for content creators to stay adaptable to navigate the ever-evolving digital landscape effectively.

Q: 3   What role does audience engagement play in authentic content creation, as described in Extract B?

- A. It is unnecessary for maintaining authenticity
- B. It serves as a means to manipulate audience perceptions
- ✓ Correct Ans **C. It demonstrates a genuine interest in the audience's feedback**
- D. It creates barriers between creators and their audience

Explanation

Extract B highlights that authentic content creators genuinely engage with their audience, demonstrating a sincere interest in their feedback and opinions.

Q: 4 How does authentic content creation contribute to setting creators apart in a competitive landscape, as discussed in Extract B?

- A. By conforming to industry trends and standards
- B. By producing content solely for financial gain
- ✓ Correct Ans **C. By embracing a unique voice and perspective**
- D. By imitating the content of other successful creators

Explanation

Extract B suggests that authenticity allows creators to stand out by embracing their unique voice and perspective amidst competition.

Q: 5 In Extract A, what is emphasized as a fundamental aspect of successful content creation?

- A. The utilization of advanced navigation technologies
- B. The reliance on traditional storytelling methods
- ✓ Correct Ans **C. The importance of adapting to evolving platforms**
- D. The integration of real-time audience feedback

Explanation

Extract A highlights the necessity for content creators to adapt to evolving platforms and technologies to maintain success.

Q: 6 What role do marine biologists and scientists play in protecting the Red Sea, according to Extract B?

- A. Advocating for sustainable tourism practices
- ✓ Correct Ans **B. Monitoring and researching the Red Sea's ecosystems**
- C. Enforcing fishing quotas and pollution control measures
- D. Organizing outreach programs for local communities

Explanation

Extract B highlights the role of marine biologists and scientists in monitoring and researching the Red Sea's delicate ecosystems to inform conservation strategies.

Q: 7 According to Extract A, what threatens the beauty of the Red Sea's delicate ecosystems?

✓ Correct Ans **A.** Overfishing and pollution
B. Climate change and rising sea temperatures
C. Coastal development and tourism
D. Industrial activities and shipping

Explanation

Extract A highlights overfishing and pollution as threats to the Red Sea's delicate ecosystems, endangering its beauty and biodiversity.

Q: 8 What is the primary message conveyed in Extract B regarding sustainable tourism?

A. Tourism operators should prioritize profit over environmental concerns.
B. Tourists should avoid visiting the Red Sea to minimize their impact.
✓ Correct Ans **C.** Tourism can contribute positively to conservation efforts when managed responsibly.
D. Sustainable tourism practices have no significant impact on the Red Sea's ecosystems.

Explanation

Extract B suggests that sustainable tourism practices, when managed responsibly, can contribute positively to the conservation of the Red Sea's ecosystems.

Q: 9 Which aspect of the Red Sea's beauty is emphasized in Extract A?

A. Its role as a sanctuary for marine life
✓ Correct Ans **B.** Its tranquility and serenity away from modern life
C. Its vibrant coral reefs and diverse marine species
D. Its geological history and ancient civilizations

Explanation

Extract A emphasizes the tranquility and serenity of the Red Sea away from modern life, portraying it as a sanctuary for those seeking solace and connection with nature.

Q: 10 What distinguishes the Red Sea as a marine marvel, according to Extract A?

A. Its proximity to African and Arabian continents
B. Its vulnerability to human activities
✓ Correct Ans **C.** Its unique geological history and biodiversity
D. Its popularity as a tourist destination

Explanation

Extract A distinguishes the Red Sea as a marine marvel due to its unique geological history and biodiversity, setting it apart from other marine ecosystems.

Q: 11 How does Extract B characterize the role of tourism operators in protecting the Red Sea's ecosystems?

- A. Prioritizing environmental concerns over profit
- B. Avoiding tourism activities altogether
- ✓ Correct Ans C. Promoting sustainable tourism practices
- D. Disregarding the impact of tourism on the environment

Explanation

Extract B characterizes the role of tourism operators in protecting the Red Sea's ecosystems by promoting sustainable tourism practices that minimize negative impacts on the environment.

Q: 12 According to Extract A, what contributes to the allure of the Red Sea?

- A. Its vulnerability to climate change
- B. Its popularity as a tourist destination
- ✓ Correct Ans C. Its tranquil and serene environment
- D. Its proximity to African and Arabian continents

Explanation

Extract A suggests that the Red Sea's tranquil and serene environment contributes to its allure, offering a sanctuary away from the hustle and bustle of modern life.

Q: 13 What is the implied meaning of the phrase "How sleep the brave"?

- A. The bravery of soldiers during sleep
- ✓ Correct Ans B. The peaceful rest of courageous individuals
- C. The lack of bravery in sleeping heroes
- D. The bravery of those who never sleep

Explanation

The phrase suggests that the brave soldiers, now deceased, rest peacefully.

Q: 14 Which natural element is personified in the poem?

- A. Wind
- B. Rain
- ✓ Correct Ans C. Spring
- D. Sun

Explanation

Spring is personified with the description of having "dewy fingers cold," attributing human-like qualities to the season.

Q: 15 What does the line "She there shall dress a sweeter sod" suggest?

- A. The beauty of a sunny day in spring
- ✓ Correct Ans B. The transformation of the burial ground by nature
- C. The arrival of fairies to adorn the graves
- D. The mourning of unseen forms

Explanation

The line implies that Spring will beautify the resting place of the fallen soldiers by transforming it into a more pleasant and peaceful space.

Q: 16 What is the role of Honour in the poem?

- A. Singing a dirge for the fallen soldiers
- ✓ Correct Ans B. Blessing the clay that covers the heroes' graves
- C. Personifying the beauty of nature
- D. Decorating the graves with flowers

Explanation

Honour is depicted as a pilgrim blessing the turf that covers the soldiers' remains, showing respect and reverence for their sacrifice.

Q: 17 Which emotion is attributed to Freedom in the final stanza?

- A. Joy
- ✓ Correct Ans B. Sorrow
- C. Anger
- D. Apathy

Explanation

Freedom is depicted as a weeping hermit, suggesting a sense of sadness or mourning over the fallen heroes.

Q: 18 Which one fits in (1) ?

- A. Sentence A
- B. Sentence B
- ✓ Correct Ans C. Sentence G
- D. Sentence H

Explanation

The correct answer is "C" it gives an introduction about the industrialization.

Q: 19 Which one fits in (2) ?

✓ Correct Ans **A.** Sentence A
B. Sentence B
C. Sentence C
D. Sentence D

Explanation

The correct answer is "A" because the agricultural production is explained in this paragraph.

?

Q: 20 Which one fits in (3) ?

A. Sentence A
B. Sentence D
✓ Correct Ans **C.** Sentence B
D. Sentence E

Explanation

The correct answer is "C" because the sentence before explains about the production of goods and the correct answer should support that line.

Q: 21 Which one fits in (4) ?

A. Sentence A
B. Sentence B
C. Sentence H
✓ Correct Ans **D.** Sentence E

Explanation

The paragraph should be linked with urbanization. Thus, the correct answer should support that idea

Q: 22 Which one fits in (5) ?

A. Sentence A
B. Sentence B
✓ Correct Ans **C.** Sentence C
D. Sentence D

Explanation

The first part of the paragraph depicts the technological advancements. Thus, the correct answer also contains that word.

Q: 23 Which one fits in (6) ?

- A. Sentence A
- B. Sentence F
- C. Sentence C
- ✓ Correct Ans D. Sentence D

Explanation

The correct answer should be related to the prior line.

Q: 24 Which one fits in (7) ?

- A. Sentence A
- B. Sentence G
- ✓ Correct Ans C. Sentence H
- D. Sentence C

Explanation

The correct answer is 'C,' because the sentence should serve as a concluding note to the whole essay.

Q: 25 In the poem, what does the speaker yearn for in the music?

- A. Loud and energetic melodies
- ✓ Correct Ans B. Slow and calming melodies
- C. Harsh and jarring melodies
- D. Chaotic and dissonant melodies

Explanation

The speaker yearns for slow and calming melodies that can bring rest and healing.

Q: 26 Which word in the poem emphasizes the sense of touch?

- A. Melody
- ✓ Correct Ans B. Fingertips
- C. Head
- D. Rhythm

Explanation

The word "fingertips" emphasizes the sense of touch as the speaker longs for music to flow over them.

Q: 27 What is the deeper meaning conveyed by the line, "A spell of rest, and quiet breath, and cool"?

- A. The speaker's desire for physical comfort
- ✓ Correct Ans B. The calming and healing power of music
- C. The presence of an actual spell in the poem
- D. The speaker's wish for a colder climate

Explanation

The line "A spell of rest, and quiet breath, and cool" conveys the deeper meaning that music has the power to provide rest, calmness, and healing to the speaker's troubled soul.

Q: 28 Which of the extracts discusses opening up a whole new world of perspectives, cultures, and traditions?

- A. Extract A
- B. Extract B
- ✓ Correct Ans C. Extract C
- D. Extract D

Explanation

Extract C states that having friends opens up a whole new world of perspectives, cultures, and traditions.

Q: 29 Which of the extracts refers to the glue that hold communities together?

- A. Extract A
- B. Extract B
- C. Extract C
- ✓ Correct Ans D. Extract D

Explanation

According to Extract D, companionship is the glue that holds communities together, fostering unity and empathy.

Q: 30 Which of the extracts depicts the bond that makes the challenges more manageable?

- A. Extract A
- ✓ Correct Ans B. Extract B
- C. Extract C
- D. Extract D

Explanation

Extract B mentions that with helpful co-workers by your side, challenges become more manageable, and achievements become even more rewarding.

Preptive Prepare Better
Series I Mathematical Reasoning Trial Test

Answers and Solutions

Q: 1 Rishi drew lines to form triangles and stars.

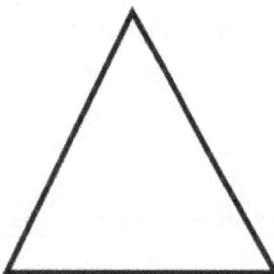 

Rishi formed a total of 10 triangles and stars. He drew 48 more lines for the stars than for the triangles. How many stars did he form?

- A. 10 stars
- B. 48 stars
- ✓ Correct Ans C. 6 stars
- D. 5 stars

Explanation

Rishi formed 'n' stars and '10-n' triangles.

Lines for stars = nx10
Lines for triangles = (10-n)x3 = 30 - nx3

Difference = nx10 - (30 - nx3) = nx13 - 30 = 48
nx13 = 78 ; n = 6

Q: 2 There are some \$2 and \$5 notes in a box.
There are thrice as many \$2 notes as \$5 notes.
Given that the total amount of money in the box is \$2255, how many \$2 notes are there in the box?

- A. 205 notes
- B. 612 notes
- ✓ Correct Ans C. 615 notes
- D. 2255 notes

Explanation

There are 3n \$2 notes and n \$5 notes.
Total value = 3n x 2 + n x 5 = n x 11 = 2255
n = 205; 3n = 615

Q: 3 Yash bought sandwiches at the prices shown below.

Salmon Sandwiches

$ 9.60 each

Tuna Sandwiches

$ 6.40 each

Yash spent $464 on some salmon and tuna sandwiches. He bought 15 more salmon than tuna sandwiches. How many tuna sandwiches did Yash buy?

- A. 320
- ✓ Correct Ans B. 20
- C. 464
- D. 960

Explanation

Yash bought 'n' tuna and 'n+15' salmon sandwiches.
Cost of tuna sandwiches = 6.40 x n
Cost of salmon sandwiches = 9.60 x (n+15) = 9.60 x n + 144

Total = 6.40 x n + 9.60 x n + 144 = 464; nx16 = 464 - 144 = 320
n = 20

Q: 4 A box of pies costs $15 while a box of muffins costs $17. Each box contained either 4 pies or 3 muffins. Mrs Silva bought the same number of pies and muffins. She paid $207 more for the muffins.
How many boxes of pies did she buy?

- ✓ Correct Ans A. 27
- B. 9
- C. 23
- D. 45

Explanation

Assume 12 pies, 12 muffins in a group.
Pies = 3 boxes, Muffin = 4 boxes
Pie box cost = $45
Muffin box cost = $68
Difference = $68 - $45 = $23
For $23 difference, 3 boxes of pies.
For a $207 difference, there are 27 boxes of pies.

Q: 5 The table shows the charges for bicycle rental.

Bicycle rental charges	
For the first hour	$8
For every additional $\frac{1}{2}$ hour or part thereof	$1.10

Shyam rented a bicycle from 8.30 a.m. to 11.45 a.m. How much did he pay for the rental of the bicycle?

✓ Correct Ans **A.** $13.50

B. $5.50

C. $58

D. $3.15

Explanation

8.30 am → 11.45 am = 3hr 15 min
1st hour = $8
Next 2hr 15 min = $1.10 x 5 = $5.50
$5.50 + $8 = $13.50

Q: 6   Mrs Tan had two rolls of ribbons of the same length but different designs. She cut the first roll of ribbon into equal pieces of length 40 cm and there were 7 suns on each piece of ribbon as shown below.

First roll of ribbon

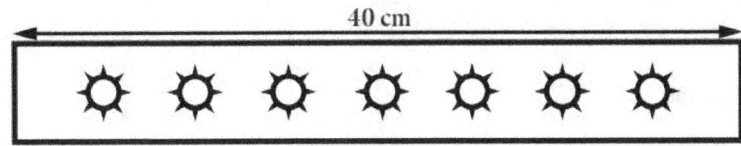

She then cut the second roll of ribbon into equal pieces of length 60 cm and there were 9 stars on each piece of ribbon as shown below.

Second roll of ribbon

After she finished cutting both rolls of ribbons, she counted that the total number of suns was 126 more than the total number of stars. Find the length of one roll of ribbon.

- A. 555 cm
- B. 120 cm
- C. 3 cm
- D. 5040 cm ✓ Correct Ans

Explanation

Common multiple = 120
Suns in 120cm = 7 x 3 = 21

Stars in 120cm = 9 x 2 = 18
1 set diff = 21 - 18 = 3

No. of sets = 126 ÷ 3 = 42
One roll = 42 x 120 = 5040
The length is 5040 cm.

Q: 7 Henry gave Zoe $420.
He then spent $\frac{1}{4}$ of his remaining money on a bag.
As a result, he had $\frac{6}{11}$ of his money left.
How much money did Henry have at first?

   A. $1440
   B. $1450
✓ Correct Ans C. $1540
   D. $1550

Explanation

Henry had 'n' at first.
After giving to Zoe, left = n - 420
After bag, left = $\frac{3}{4}$ (n - 420)

$\frac{3}{4}$ (n - 420) = $\frac{6}{11}$ n
33 x (n-420) = 24n
33n - 33x420 = 24n

9 n = 33x420 ; n = 1540

Q: 8 Connie paid $27 for a belt after a discount at a bazaar. She then bought a skirt from another shop. She spent a total of $72 on these two items altogether.
She saved $12 total from discounts.
Given that Connie was given a 10% discount on the belt, find the percentage discount given for the skirt.
Round off your answer to one decimal place if necessary.

   A. 17.5%
   B. 17.6%
✓ Correct Ans C. 16.7%
   D. 16.5%

Explanation

	Original price	Discount	Discounted price
Belt	$30	$3 i.e 10%	$27
Skirt	$54	$9 (total discount - Belt discount)	$45

Discount for Skirt = $\frac{9}{54}$ x 100% = 16.7 %

Q: 9 An apple juice factory's cylindrical tank is filled up before being transferred into a tanker. The graph below shows how the height of the apple juice in the tank changes.
How long did it take to empty the tank?

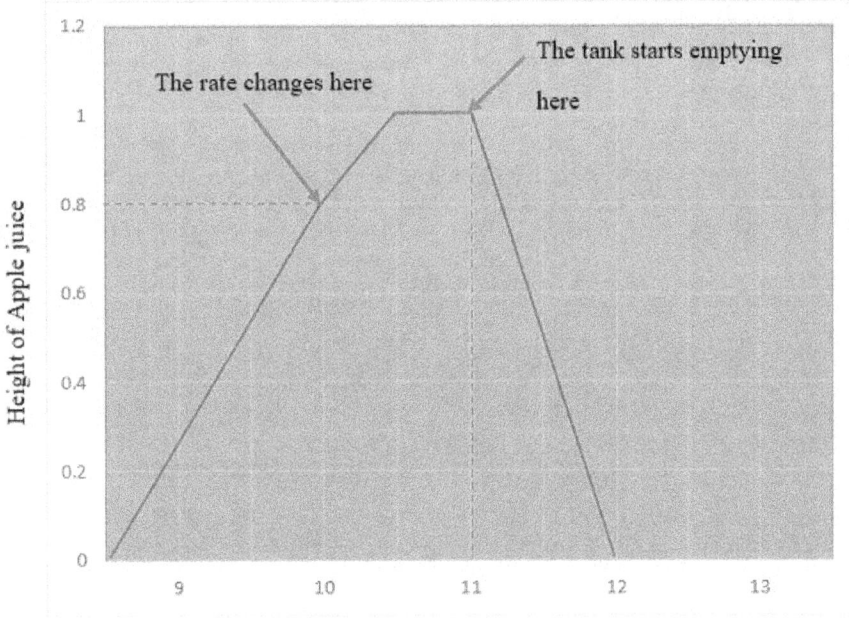

A. 1.5 hours

B. 2.5 hours

C. 2 hours

✓ Correct Ans D. 1 hour

Explanation

The downhill part of the graph shows the tank being emptied.
This starts at 11.00 and ends at 12.00 which is 1 hour.

Q: 10 The stem and leaf diagram on the right shows the ages of some school teachers. How many teachers are in their forties?

```
3 | 3  5
4 | 0  5  7  8
5 | 1  4  9
6 | 1  3
Key: 5 | 4 = 54 years
```

    A. 2

✓ Correct Ans B. 4

    C. 6

    D. 3

Explanation

The key tells you that 4 in the stem means 40. You need to count the leaves in the second row. The leaves are 4. So, 4 teachers are in their forties.

Q: 11 A final mark for a grading examination is calculated from three components using the following formula:
Component A × 0.6 + Component B × 0.3 + Component C × 0.1.
What was this candidate's final approximate mark if the candidate obtained the following marks:

- Component A = 64
- Component B = 36
- Component C = 40

    A. 76

    B. 28

    C. 64

✓ Correct Ans D. 53

Explanation

Substitute the values given for each mark in the formula.
The final mark is (Component A × 0.6 + Component B × 0.3 + Component C × 0.1)
Which is (component 64 × 0.6) +(component 36 × 0.3) + (component 40 × 0.1)
The results are 38.4 + 10.8 + 4 = 53.2 which is nearer to 53.

Q: 12 Four schools had the following proportion of pupils with special education needs. Which school had the lowest proportion of pupils with special education needs?

School	Proportion
P	2/9
Q	0.17
R	57 out of 300
S	18%

✓ Correct Ans **A.** School Q

B. School R

C. School S

D. School P

Explanation

Change each figure into decimals. $\frac{2}{9}$ = 0.222, 57 out of 300 = 0.19 and 18% = 0.18. Among the given ratios and decimals, the lowest value is 0.17. Thus, school Q has the lowest proportion.

Q: 13 Elena bought many types of toys at an average cost of $12. She then bought one of each of the following two toys and the average cost of all her toys became $14.

$28

$10

How many toys did she buy in total?

✓ Correct Ans **A.** 7

B. 5

C. 6

D. 8

Explanation

Initial total cost = 12 x n
Final total cost = 14 x (n+2)
The difference is 38.

14xn + 28 - 12xn = 38; nx2 = 10 ; n = 5
Final number of toys = n + 2 = 7

Q: 14 The diagram below shows a bar chart.

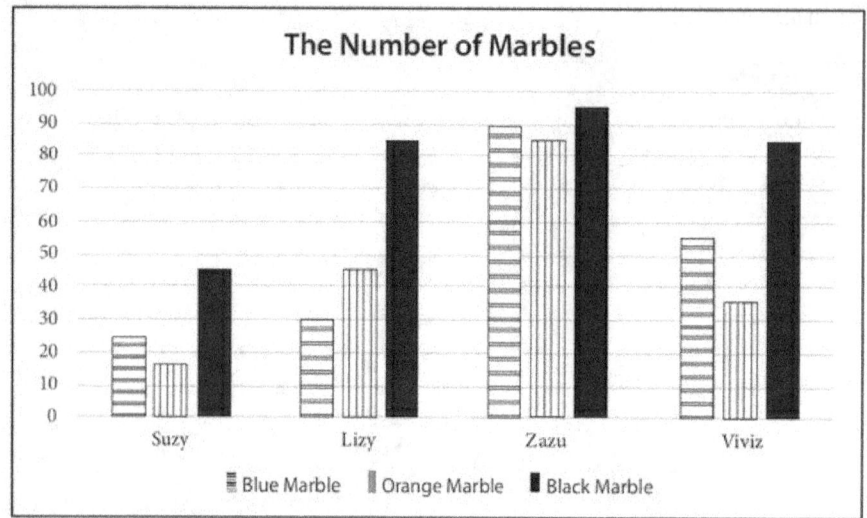

Which of the following is correct?

A. The total number of blue marbles that Viviz and Lizy have is higher than the number of blue marbles that Zazu has.

B. The total number of marbles that Suzy has is higher than the total number of Viviz's blue and grey marbles.

C. Lizy has more marbles than Viviz.

✓ Correct Ans D. Zazu has the highest number of marbles.

Explanation

A. The total number of blue marbles that Viviz and Lizy have is higher than the number of blue marbles that Zazu has. Wrong. Viviz and Lizy, 55 + 30 = 85 < zazu, 90.
B. The total number of marbles that Suzy has is higher than the total number of Viviz's blue and grey marbles. Wrong. S = 25+15+45=85, V=55+85=140.
C. Lizy has more marbles than Viviz. Wrong. L=30+45+85=160, V=55+35+85=175.
D. Zazu has the highest number of marbles. Correct. 90+85+95=270

?

Q: 15 In a trunk there are 5 chests; in each chest, there are 3 boxes; and in each box, there are 10 gold coins. The trunk, the chests, and the boxes are locked.

At least how many locks need to be opened to take out 50 coins?

A. 6

✓ Correct Ans B. 8

C. 7

D. 9

Explanation

To take out 50 coins, 5 boxes must be opened.
To open 5 boxes, 2 chests must be opened.
To open 2 chests, the trunk must be opened.
So altogether, the trunk, 2 chests, and 5 boxes must be opened = 8 locks

Q: 16 The table shows the number of beads in a container.

Colour	Number of beads
White	150
Red	130
Blue	$\frac{2}{5}$ of the number of white beads

Daisy uses 3/5 white beads and 20% of red beads to embroider her dress. Count the number of beads left in the container.

A. 225
B. 284
✓ Correct Ans C. 224
D. 275

Explanation

Blue = 2/5 x 150 = 60
White used = 3/5 x 150 = 90, balance 60
Red used = 130 x 20% = 26, balance 104
Remaining: 104 + 60 + 60 = 224

Q: 17 The diagram below shows 2 shapes.

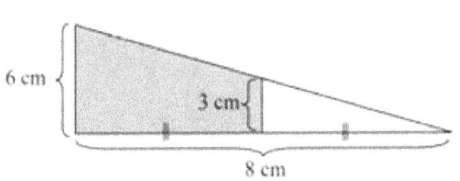 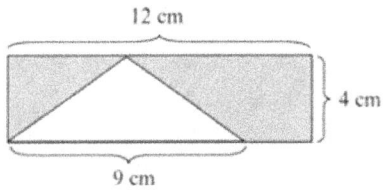

Find the total **non-shaded** area.

A. 35cm^2
✓ Correct Ans B. 24cm^2
C. 25cm^2
D. 15cm^2

Explanation

½ x 3 x 4 = 6 cm^2
½ x 4 x 9 = 18 cm^2
Total = 24cm^2

Q: 18   The diagram below shows measuring tools.

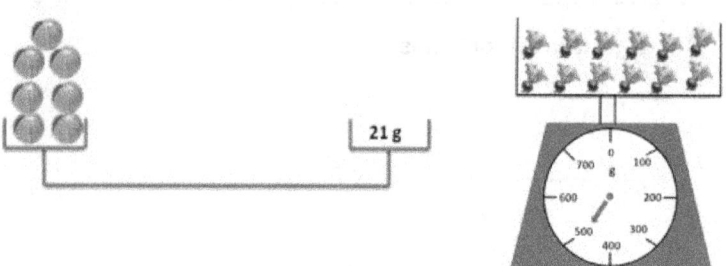

Rara buys 12 balls and 3 times beetroot shown in the picture above. How much is the total weight of all the items?

A. 1,055g

✓ Correct Ans B. 1,536g

C. 1,533g

D. 1,305g

Explanation

7 balls weigh 21 g → 1 ball = 3 g; 3 × 12 = 36 g.
500g x 3 = 1,500g
Total: 1,536g

Q: 19   The diagram below shows measuring tools.

Fifi buys 8 apples. Find the weight of the apples.

✓ Correct Ans A. 160g

B. 120g

C. 150g

D. 130g

Explanation

Find the weight of one ball. 16g ÷ 8 = 2
4 balls x 2 = 8g
28g – 8g = 20g for 1 apple.
20 x 8 = 160g

Q: 20 The diagram below shows an equation.

What is the answer?

- A. 5
- ✓ Correct Ans **B.** 7
- C. 6
- D. 8

Explanation

Let us change the pictures into letters.
R + R + R = Y + Y + 1
R + 2 = Y
Y=?

3R = R+2 + R+2 + 1
3R = 2R + 5
3R-2R = 5
R = 5
5 + 2 = Y; Y = 7

Q: 21 204 eggs were packed into big and small cartons.
Each big carton contained 12 eggs and each small carton contained 6 eggs. There were 8 more big cartons than small cartons. How many such big cartons were used?

- A. 143
- ✓ Correct Ans **B.** 14
- C. 24
- D. 18

Explanation

Big carton = n, Small carton = n - 8
Eggs = nx12 + (n-8)x6 = nx18 - 48 = 204
nx18 = 252; n = 14

Q: 22 The following pictograph shows the number of books sold by a company during a week. Study the pictograph carefully and answer the question given below.

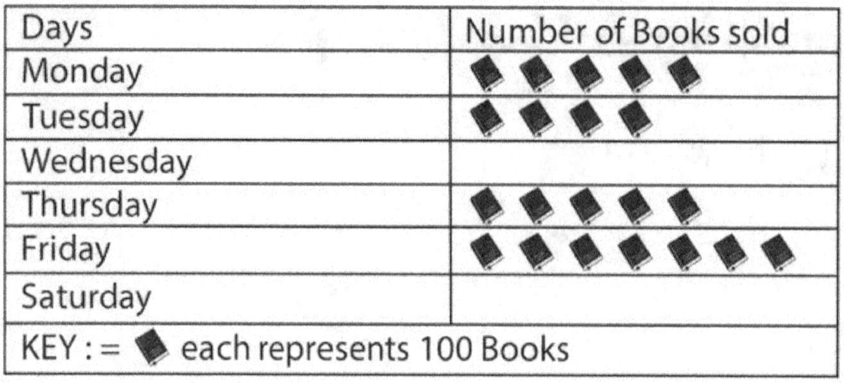

Days	Number of Books sold
Monday	📕 📕 📕 📕 📕
Tuesday	📕 📕 📕 📕
Wednesday	
Thursday	📕 📕 📕 📕 📕
Friday	📕 📕 📕 📕 📕 📕 📕
Saturday	
KEY: = 📕 each represents 100 Books	

Books sold on Wednesday are twice as many as those sold on Tuesday. Books sold on Saturday are 400 less than books sold on Wednesday. How many books were sold on **Tuesday and Saturday**?

- A. 600
- ✓ Correct Ans B. 800
- C. 260
- D. 300

Explanation

Wednesday = 4 x 2 = 8 x 100 = 800
Saturday = 800 − 400 = 400
Saturday and Tuesday = 400 + 400 = 800

Q: 23 Marlene was preparing for a race. She ran the same distance every day for five days. The graph shows the time taken, in minutes, to complete her run each day.

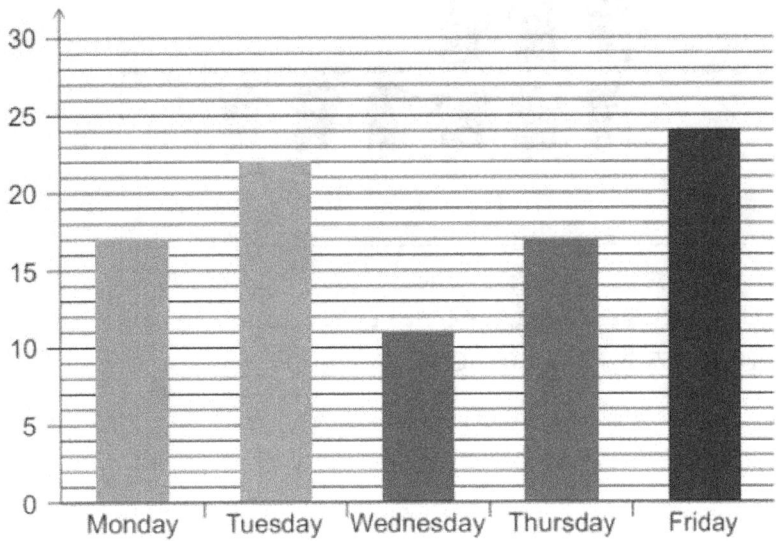

Which of these statement(s) is/are correct?

1. Marlene ran the fastest on Friday.
2. Marlene's time on Tuesday was exactly double of her time on Wednesday.
3. Marlene's total time running on Tuesday and Wednesday was the same as her total time running on Monday and Thursday.

- **A.** none of them
- **B.** Statement 1 only
- ✓ Correct Ans **C.** Statement 2 only
- **D.** Statement 3 only

Explanation

1. Marlene ran the fastest on Friday. Wrong. The fastest has the shortest time. So, the fastest is on Wednesday.
2. Marlene's time on Tuesday was exactly double of her time on Wednesday. Correct. 22/2 = 11
3. Marlene's total time running on Tuesday and Wednesday was the same as her total time running on Monday and Thursday. Wrong. 33 minutes / 34 minutes.

Q: 24 The picture shows the exact number of shirts sold every day.

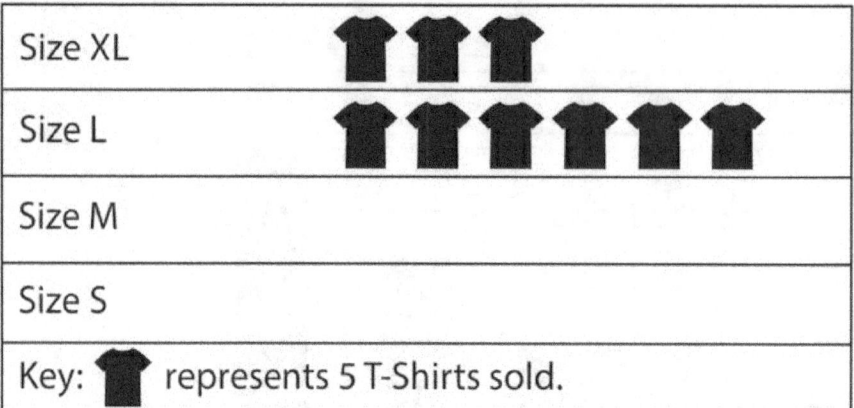

Shirts size S sold 2 times size XL, while shirt size M sold 3 times size S.
How many shirts were sold on **4 days**?

✓ Correct Ans **A.** 660

B. 550

C. 33

D. 132

Explanation

Size S: 3 x 2 = 6 , size M: 6 x 3 = 18
All shirts sold in one day: 6 + 18 + 6 + 3 = 33 symbols of shirts

33 x 4 days = 132 x 5 = 660 T-Shirts sold

Q: 25 Lisa left her house at 8:30 AM and arrived at her friend's house at 10:15 AM. If the distance between their houses is 12 kilometers, what was Lisa's **approximate** average speed during her journey?

A. 5 km/h

B. 6 km/h

✓ Correct Ans **C.** 7 km/h

D. 8 km/h

Explanation

Time = 1 hr 45 mins = 1.75 hrs
Journey = 12 kms
Speed = 12/1.75 = 12 divided by $7/4$
= $48/7$ = almost 7
Rounding to the nearest whole number, we get an average speed of approximately 7 km/h.

Q: 26 The pictograph shows the number of A.C. sets sold in several years.

Year	Number of A.C sets sold
1995	▮ ▮
2000	▮ ▮ ▮ ▮
2005	▮ ▮ ▮
2010	▮
2017	▮ ▮ ▮ ▮ ▮
KEY : Each  represents 3000 A.C sets	

Which of the following is **wrong**?

✓ Correct Ans **A.** The difference between the number of A.C. sold in 2017 and 2000 is 2,500.

B. The total number of A.C. sets sold in 2010 and 1995 is 9,000.

C. The total of the number of A.C sets sold in 2017 and 2005 is 24,000

D. None of the above.

Explanation

A. The difference between the number of A.C. sold in 2017 and 2000 is 2,500. Wrong statement but the correct answer. 5-4 = 1 x 3,000 = 3,000.
B. The total of the number of A.C. sets sold in 2010 and 1995 is 9,000. Correct statement but the wrong answer.
C. The total of the number of A.C. sets sold in 2017 and 2005 is 24,000. Correct statement but the wrong answer. 8 x 3,000 = 24,000
D. Option A. Correct statement but wrong answer.

Q: 27 Find the area of the following figure. Assume all triangles have the same dimensions.

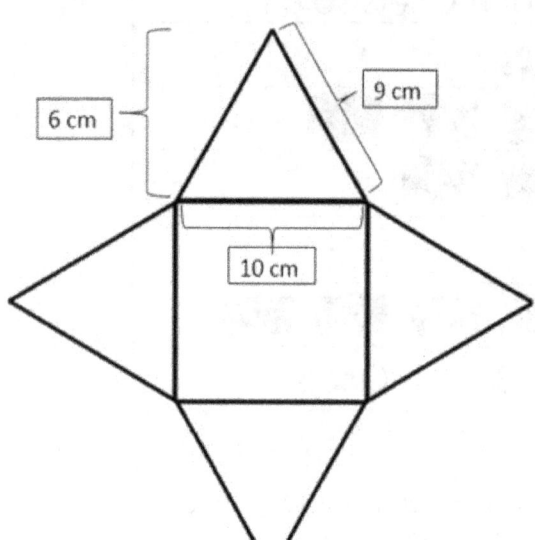

- A. 130 cm²
- B. 145 cm²
- ✓ Correct Ans C. 220 cm²
- D. 280 cm²

Explanation

Area = area of 4 triangles+ area of the square
= 4 ×1/2 ×10 ×6 +10 ×10 = 220 cm²

Q: 28

1 Tulip and a Rose cost $18
2 Roses and a Dahlia cost $25
3 Tulips and 2 Dahlia cost $34

What is the total value of 1 Tulip, 1 Dalia and 1 Rose?

A. $20
B. $23 ✓ Correct Ans
C. $25
D. $27

Explanation

Tulip=T, Rose= R, Dahlia=D
T+R=18----- (1)
2R+D=25----- (2)
3T+2D=34----- (3)

(2) ×2 - (3) :
4R+2D-3T-2D=50-34

4R-3T=16----- (4)

(1)×3+(4) :
3T+3R+4R-3T=54-16
7R=70; R=10

Substitute R=10 in (1)
T+10=18; T=8

Substitute R=10 in (2)
2 ×10+D=25 ; D=5

Total= 10+8+5 =23

Q: 29

Machine P prints 40 identical posters per hour while Machine Q prints 90 such posters per hour. Machine P started printing 30 minutes before Machine Q. When the number of such posters printed by Machine P was equal to the number of such posters printed by Machine Q, both machines were switched off to stop the printing.

What was the total number of copies printed by the two machines?

A. 72 ✓ Correct Ans
B. 42
C. 36
D. 76

Explanation

In 'n' hour after starting together, both machines have same number of posters printed.

Machine P prints = 20 + 40n
Machine Q prints = 90n
Total printed = 130n + 20

20 + 40n = 90n; 50n = 20; n = $2/5$
Total = 130n + 20 = 130 x $2/5$ + 20 = 72

Q: 30 The average of four **3-digit numbers** is 500.
Two of the numbers are 150 and 230.
What is the **largest difference** between the other two numbers?

　　A. 348

✓ Correct Ans B. 378

　　C. 358

　　D. 366

Explanation

Sum of four numbers → 500 × 4 = 2,000
Sum of other 2 numbers → 2000 − 150 − 230 = 1,620
Largest number possible → 999
Smaller number → 1,620 − 999 = 621
Largest difference → 999 − 621 = 378

Q: 31 Connie started saving money in her piggy bank on a Friday. She saved $1 per day from Monday to Friday and $2 per day on Saturday and Sunday.
On which day of the week would she have saved $32 in her piggy bank?

✓ Correct Ans A. Sunday

　　B. Tuesday

　　C. Friday

　　D. Saturday

Explanation

(5 × $1) + (2 × $2) = $9
$32 ÷ $9 = 3 R $5
3 groups of Fri, Sat, Sun, Mon, Tue, Wed, and Thu.
Remaining $5 → Fri ($1), Sat ($2), and Sun ($2).
Connie would have saved a total of $32 in her piggy bank on Sunday

Q: 32 Class X has 50% more students than Class Y. Class X has 40% fewer students than Class Z. Given that there are 80 students in the 3 classes, how many students are there in Class X?

A. 22
B. 24 ✓ Correct Ans
C. 28
D. 32

Explanation

Class Y = n
Class X = nx1.5

0.6 x Class Z = nx1.5
Class Z = nx1.5/0.6 = nx2.5

Total = n + nx1.5 + nx2.5 = 5n = 80
n = 16
Class X = nx1.5 = 16x1.5 = 24

Q: 33

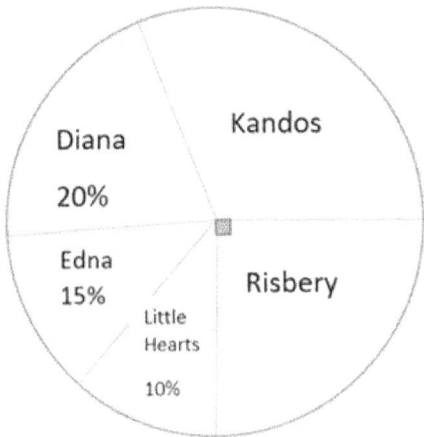

This chart shows favourite chocolate brands of 150 students.
Answer question using the above chart.

How many students like Kandos?

A. 30
B. 45 ✓ Correct Ans
C. 55
D. Cannot be said

Explanation

Risbery has got 90° of the pie chart. 90° is $1/4^{th}$ of the total pie chart (360°). So it must get $1/4^{th}$ of the total percentage. So the percentage for Risbery is 25%. Remaining percentage is for Kandos.

Percentage for Kandos= 100-(20+15+10+25) = 100 - 70 = 30%
No. of students who like Kandos= 150 x30/100 = 45

Q: 34

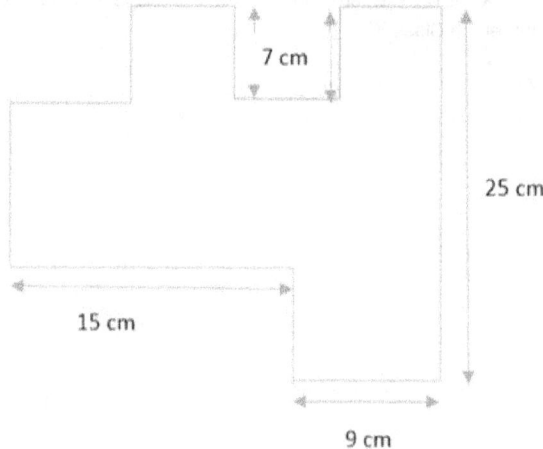

What is the perimeter of the given figure?

A. 98 cm
B. 100 cm
✓ Correct Ans C. 112 cm
D. Data is not enough

Explanation

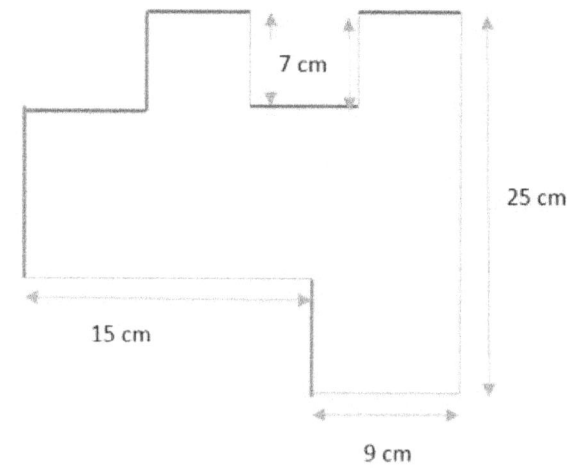

Highlighted vertical lines must add up to 25 cm.
Highlighted horizontal lines must add up to 15+9 = 24 cm

Perimeter = 25+25+24+24+7+7 = 112 cm

Q: 35 The table shows the number of newspapers sold by Mr. Johan.

English Book	Number of sales
January	900
February	60 % more than the number of sales in January
March	20% lower than the number of sales in February
April	700

Count the number of newspapers sold by Mr. Johan in 4 months total.

- **A.** 4,850
- **B.** 4,459
- **C.** 4,561
- ✓ Correct Ans **D.** 4,192

Explanation

February :
900 x 60% = 540
900 + 540 = 1,440

March :
1,440 x 20% = 288
1,440 − 288 = 1,152

Total = 1,152 + 1,440 + 700 + 900 = 4,192

Preptive Prepare Better
Series I Thinking Skill Trial Test

Answers and Solutions

Q: 1 When Georgia was 10 years old, her parents promised to buy her a pet. That's because pets aren't allowed in the apartment complex they live in. Fortunately, they had some plans to relocate within a year. If she is 12 years old now, which of the following statements is possible?

  I. She has a pet.
  II. She doesn't have a pet.

  A. Statement I
  B. Statement II
  ✓ Correct Ans C. Both Statements
  D. None of the above

Explanation

Both statements are possible since she could only get a pet once they relocated. If they left that apartment building, there is a high chance she has a pet. Similarly, if they are still in that complex, Georgia doesn't have a pet.

Q: 2 If you read one book by this author, there is a high chance of considering reading more of his work. That's according to research that shows 98% of people who have read one of his books have reacted in that manner. His way of captivating his readers is responsible for this behavior.

James: "Anyone who has read a book from this author has read more than one of them."
Graham: "One of the keys to get a huge fan base as an author is to captivate your readers."

Which of the following observations is correct?

  A. James Only
  ✓ Correct Ans B. Graham Only
  C. Both James and Graham
  D. Neither James nor Graham

Explanation

James is wrong since there is the 2% that doesn't read a second book, according to the quoted research. Graham is right, going by the last statement.

Q: 3 Use the following statements to answer the question.

- Hunter will be visiting his grandmother after the school closes this weekend.
- He will only see his grandmother if his covid tests come back negative.
- Hunter hasn't seen his grandmother for years now.

After 2 weeks :

Rue: "The results were negative for Hunter since I saw a very recent photo of Hunter and his grandmother."
Jenna: "Schools have closed recently since Hunter's trip to her grandmother's house was scheduled one week back."

Which of the following reasoning is correct?

- **A.** Rue Only
- **B.** Jenna Only
- ✓ Correct Ans **C.** Both Rue and Jenna
- **D.** Neither Rue nor Jenna

Explanation

Given the second and third statements, Rue is right. Jenna is also correct, given the first statement.

Q: 4 A dice is numbered from 1 to 6 in different ways.
If 1 is adjacent to 6, 2, or 4, which of the following statements is definitely correct?

- **A.** 3 is opposite to 5
- ✓ Correct Ans **B.** 3 is adjacent to 5
- **C.** 1 is adjacent to 3
- **D.** 2 is opposite to 6

Explanation

Given the sides that 1 is adjacent to, it can only lie opposite to 5 or 3. Consequently, the possible pair can't lie opposite to one another. So, they can only be adjacent; hence, B is correct.

Q: 5 If you regularly hike in nature, going to the treadmill is unnecessary. Equally important, you don't need high-tech gadgets to exercise. The only thing necessary is a comfortable pair of hiking boots.

What is the best interpretation of the above statements?

- **A.** Hiking is better than using the treadmill for overall fitness.
- **B.** Using comfortable hiking boots is crucial for a successful workout.
- ✓ Correct Ans **C.** Hiking regularly is as effective as using the treadmill for fitness.
- **D.** Combining hiking and treadmill workouts is essential for a balanced exercise routine.

Explanation

The paragraph suggests that hiking and treadmill workouts are equivalent exercises; hence, A and D are incorrect, whereas C is the correct interpretation. The mention of comfortable hiking boots is more of a recommendation for comfort, so B is not the main point.

Q: 6 Different people had various opinions regarding the price of a laptop. A said that the price was between $800 and $900. B had a different range of between $750 and $850. C said that the price is less than or equal to $840.

If all the opinions were factual, which of the following conclusions is correct?

- A. The price can be $860
- B. The price can also be $920
- ✓ Correct Ans C. The price can be $820
- D. The price can be $880

Explanation

Only C satisfies all the criteria.

Q: 7 In the music competition that started nine years ago, no participant has won the first place twice. Last season, Sophia claimed the victory, while Benjamin finished as the runner-up. According to this year's result, Sophia didn't break the tradition despite competing again. Benjamin also took another shot at the competition but did even worse than before. The only other contestants were Emma and Alexander.

If Alexander ended up in the last position, who won this time?

- A. Sophia
- B. Benjamin
- ✓ Correct Ans C. Emma
- D. Alexander

Explanation

Since Alexander was the last, he can't be the winner; hence, D is incorrect. Sophia won last year's contest but didn't break the tradition of winning the first place once. In other words, she didn't win this time either; hence, A is also incorrect. Benjamin was the runner-up last time, and since he performed even poorer, he can't be the winner, which means B is also incorrect. Therefore, the winner was Emma; hence, C is the correct answer.

Q: 8 Use the following statements to answer the following question.

I. Elsa is 49 years old.
II. Nelson is her age mate.
III. Elsa has brown hair.
IV. Nelson has the same hair color.
V. Elsa has long hair.

Which statements helps you know that Elsa and Nelson are identical twins?

- A. I, II, and III
- ✓ Correct Ans B. I, II, III, and IV
- C. All the above
- D. None of the above

Explanation

Out of all the statements, only the first four can help you know that Elsa and Nelson are twins, because the characteristics are described in pairs.

Q: 9 A detective investigates a series of art thefts across Europe. The suspect leaves cryptic clues at each crime scene, each consisting of a geometric shape and a number. At the Louvre, the clue is a triangle and the number 10. At the Rijksmuseum, the clue is a pentagon and the number 15.

Based on this pattern, what shape and number should the detective expect at the next crime scene?

- A. Square, 20
- ✓ Correct Ans B. Heptagon, 20
- C. Circle, 30
- D. Impossible to determine

Explanation

Analyze the pattern:

- The number of sides in the shape increases by 2 at each crime scene (triangle, pentagon).
- The associated number increases by 5 at each scene (10, 15).

Therefore, the next scene should have a shape with 5 + 2 = 7 sides (heptagon) and a number increased by 5 again (20).

Q: 10 Roy loves music. Whenever he is in the mood for playing his favourite music, he plays multiple tracks. Today, he selected 3 of his tracks and put them in his 3 disc players to play them. The singles are of the same duration: 3 minutes. He started playing them at once. For how long did he enjoy the music play?

- A. 12 minutes
- B. 9 minutes
- C. 60 minutes
- ✓ Correct Ans D. 3 minutes

Explanation

Roy plays all the CDs together at once. So, each CD will play for 3 minutes, as such he hears the music for 3 minutes. Option D is correct.

Q: 11 By their nature, their large number and the diversity of their operations/ sectors, family businesses are of great economic value in any economy. However, the sad news is that just a few percent of them barely survive past the third generation. Many family businesses get into problems for many reasons, including being badly affected by family conflicts. Also, weak or non-existent corporate governance, and favouritism in allocating resources, positions and appraisal are major challenges faced by family businesses.

If the given information is true, which of the following statements must be false?

- A. Family businesses are owned and run by family members.
- B. Many family businesses do not survive past the third generation.
- ✓ Correct Ans C. Family businesses do not have the challenges of weak corporate governance or authority allocation.
- D. Family conflicts most often adversely affect family businesses.

Explanation

Based on the facts given above, family businesses are badly affected by family conflicts, weak or non-existent corporate governance, favouritism in allocating resources, positions and appraisal. So C is not true.

Q: 12 Three new pizzas, code-named A, B, and C are being developed by a restaurant.
A contains mushroom, spinach, cheese, tomatoes, red pepper and onion.
B contains cheese, spinach, tomatoes, red pepper and onion.
C contains mushrooms, cheese, tomatoes, onion and red pepper.

Which ingredients are included in only two ?

- ✓ Correct Ans A. Mushroom and Spinach
- B. Cheese and Spinach
- C. Mushroom and Onions
- D. Tomatoes and Spinach

Explanation

Only A and C contain Mushrooms and only A and B contain Spinach. Other ingredients are same in all three pizzas. Option A is correct.

Q: 13 Consider the following notice at the supermarket during the Christmas week.

> **NOTICE.**
>
> The store will be opened this week on the following days and hours:
>
> - Weekdays: 9.00 am – 9.00 pm
> - Weekends: 8.00 am – 10.00 pm
> - Christmas Day: 1.00 pm to 7.00 pm
>
> Thanks for your understanding.
>
> **Management**

How many hours was the supermarket open that week, if the week starts on a Sunday and Christmas day was on Friday?

- A. 76
- B. 82 *(Correct Ans)*
- C. 88
- D. 91

Explanation

The affected days are Sunday to Saturday.
Opening hours are:

- Sunday: 8.00 am to 10.00 pm = 14 hours
- Monday - Thursday: 9.00 am to 9.00 pm = 12 hours X 4 = 48
- Christmas Day – 1.00 pm – 7.00 pm = 6 hours
- Saturday: 8.00 am – 10.00 pm = 14 hours
- Total = 82 hours

Option B is correct.

Q: 14 Trish is a famous painter.
Some of her exhibitions are packaged by her brother.

Which of the following is most likely correct, based on what we know so far?

- A. Trish has done some exhibitions as a famous painter *(Correct Ans)*
- B. Trish's brother is a businessman
- C. Trish's paintings are sold only at exhibitions
- D. Trish's brother loves promoting painting exhibitions

Explanation

Trish has done some exhibitions promoted by her brother, but there is no information if her brother is a businessman or promotes other painters' work or if her works are sold only at exhibitions.
Option A is correct.

Q: 15   The Tennis competition started on the 13th April, Saturday.
Matches are only on weekend days and Wednesdays. There are 24 preliminary matches with 4 matches each match day. There are 12 Round-Two matches, with 3 matches each match day, the two semi-final matches were played on two match days. There was a rest day (in match day) before the final match. When was the final played?

- A. May 4th
- B. May 8th
- ✓ Correct Ans C. May 12th
- D. May 5th

Explanation

Preliminary matches were 24 matches, 4 each day = 6 match days
R2 matches are 12, 3 each day = 4 match days
Semi, 2 matches, 1 each day = 2 match days
Rest day   = 1 match day
Final   = 1 match day
Total   = 14 match days

The matches were on Saturdays, Sundays and Wednesdays.
It started on April 13, Saturday
Playing days are:
Saturday –   April: 13, 20, 27, May: 4, 11
Sundays -   April: 14, 21, 28. May: 5, 12
Wednesdays – April: 17, 24, May:1,   8, 15
From April 13, Sunday, May 12th is the 24th match day.
Option C is correct

Q: 16 The following tables show the time returned by five athletes in two races.
The athlete with the maximum improvement and completing Race 2 in equal to or less than 10.8 seconds will get an award.

Who got the award ?

Race 1

Colin	11.0 sec
David	11.2 sec
Gabriel	10.9 sec
Paul	11. 3 sec
Chris	11.1 sec

Race 2

Colin	11.0 sec
David	10.9 sec
Gabriel	10.8 sec
Paul	11.1 sec
Chris	10.8 sec

✓ Correct Ans **A.** Chris

B. Paul

C. Gabriel

D. David

Explanation

The improvements over the two races are:

- Colin: 11.0 sec to 11.0 sec = no improvement
- David: 11.2 sec to 10.9 sec = 0.3 sec
- Gabriel: 10.9 sec to 10.8 sec = 0.1 sec
- Paul: 11.3 sec to 11.1 sec = 0.2 sec
- Chris: 11.1 sec to 10.8 sec = 0.3 sec

The two runners with the biggest improvements are David (0.3 sec) and Chris (0.3 sec).
However, David did not meet the qualifying time of 10.8 sec. Chris did, so he won the award. Option A is correct.

Q: 17 A library has three borrowing rules:

• You can only borrow maximum two books at a time.
• You cannot borrow fiction books and nonfiction books on the same day.
• You cannot borrow a book on the same day you return another book.

If you return a book on Monday, can you borrow a fiction book on Tuesday?

A. Yes, always.

✓ Correct Ans **B.** Yes, if you do not borrow a non-fiction book on Tuesday

C. No, never.

D. It depends on whether you borrowed a fiction book on Friday.

Explanation

The rule only prohibits borrowing on the same day as returning. Returning a book on Monday does not prevent you from borrowing another book on the following day, Tuesday, as long as the book you borrow is not subject to the "fiction and nonfiction on the same day" restriction.

Q: 18   You are trapped in a room with four doors labeled A, B, C, and D. Each door leads to a different outcome:

• Door A: You return to the start of the room.
• Door B: You teleport to a hidden chamber with another door leading back to the room.
• Door C: You escape the room and win, only if you come through another door.
• Door D: You encounter a riddle that, if solved correctly, opens Door C.

A sign tells you: "Only one door leads to freedom, others trap you further."

You can only choose one door. Which door should you choose?

   A. Door A
   B. Door B
   C. Door C
   ✓ Correct Ans D. Door D

Explanation

Door A and C (first time) are unlikely candidates based on the sign's message.
Door B might lead to another loop, so taking a risk on Door D seems wiser.
Solving the riddle on Door D offers a potential path to freedom, through C.

Q: 19   You are driving on a road with three intersections, each controlled by a traffic light. Each light can be red, yellow, or green. You know that:

• At least one light is always green.
• No two consecutive lights are green.
• If one light is red, the next light must be yellow.

What is the possible combination of colors for the three lights, in order from left to right?

   ✓ Correct Ans A. Green, yellow, red
   B. Red, green, yellow
   C. Yellow, green, red
   D. Any combination is possible

Explanation

At least one green light is required.
No two consecutive greens -> this satisfies all options.
If one light is red, the next must be yellow, eliminating red-green-yellow and yellow-green-red.

Therefore, the only remaining valid combination is green, yellow, red.

Q: 20 A computer store assigns unique identification numbers to its laptops using the following system.

- The first digit represents the brand (1 for Dell, 2 for HP, 3 for Lenovo).
- The second digit represents the storage type (1 for SSD, 2 for HDD, others unknown).
- The remaining three digits represent the laptop's model number.

If a laptop's ID number is 23145, which of the following statements must be true?

- **A.** It is an HP laptop with an HDD.
- **B.** It is the 145th model of Lenovo laptops.
- **C.** It is a Dell laptop with an SSD.
- ✓ Correct Ans **D.** It is the 145th model of HP laptops.

Explanation

The first digit (2) indicates HP, and the second digit (3) is unknown. The remaining three digits (145) represent the model number.

Q: 21 Three friends, Alice, Bob, and Carol, provide information about their weekend plans:

Alice: "I am either going to the beach or the mountains."
Bob: "If Alice goes to the beach, I'll join her. Otherwise, I'll stay home."
Carol: "I'll join Bob wherever he goes."

If Carol stays home for the weekend, where is Alice most likely to be?

- **A.** Beach
- ✓ Correct Ans **B.** Mountains
- **C.** River
- **D.** Impossible to determine

Explanation

If Carol stays home, it implies that Bob stays home as well. So Alice didn't go to beach.

Q: 22 You're hiking in a remote area and stumble upon a hidden cabin. Inside, you find a diary detailing a magician's life, ending with a cryptic entry: "*Farewell, stage. I vanish to where shadows dance.*"

Where might the magician have gone, based on the clue?

- **A.** A dark cave system shrouded in perpetual darkness.
- **B.** A secluded island surrounded by dense, sun-blocking foliage.
- **C.** A bustling city with a vibrant nightlife and hidden alleyways.
- ✓ Correct Ans **D.** An abandoned theater, their final performance space now consumed by darkness.

Explanation

The clue mentions "shadows dancing," which suggests a place associated with darkness and theatricality.
While caves, islands, and cities can have shadows, the mention of "stage" and "vanishing" directly points towards the abandoned theater, a place where the magician's art thrived and ultimately faded into darkness.

Q: 23 If the mirror is placed on the left line, then which of the given figures will be the correct image of the question figure.

A.

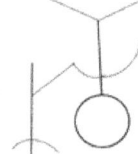

✓ Correct Ans B.

C.

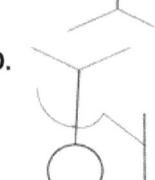

D.

Explanation

Correct image of the question figure is image B.

Q: 24 Gone are the days when Emma would comfortably go to work by bus. Since one of the buses broke down, people are scrambling for the remaining seats. Sometimes, the bus is full to capacity when it reaches the station where Emma takes her bus. Even when it isn't full, it is too crowded for Emma to enjoy her ride.

Based on the above information, what is likely true ?

  A. Emma should start taking the train instead.
  B. Emma will most likely look for another job to avoid this route.
✓ Correct Ans C. Since one of the buses broke down, Emma's commute has been inconvenienced.
  D. Many passengers are complaining about going by bus.

Explanation

The only statements that the above facts support is C since her problems started when one of the buses broke down.

Q: 25 The code of a certain language is as follows :
126 means **Mary is good.**
425@ means **Good things are amazing.**
562 means **Bring good things**.

Which of the following is a possible code for *these things are good*?

- A. 257#
- B. 1234
- C. 54#$
- ✓ Correct Ans D. 425^

Explanation

2 -> good
5 -> things
are -> @ OR 4

So the wxpected one should have 2, 5 and either @ or 4.

Q: 26 A store offers a special discount on specific items. The price tag shows the original price followed by a cryptic clue: "Subtract the product of the two digits". For example, a $40 item has the clue "75", implying $40 – (7 x 5) = $5.

You encounter three items with the following price tags and clues:

- Item A: $32, "24"
- Item B: $87, "51"
- Item C: $148, "69"

Which item has the highest discounted price after applying code ?

- A. Item A
- B. Item B
- ✓ Correct Ans C. Item C
- D. None of the item.

Explanation

Analyze the discounted prices for each item :
Item A: $32 minus the product of its first two digits (2 and 4) results in $24.
Item B: $87 minus the product of its first two digits (5 and 1) gives $82.
Item C: $148 minus the product of its first two digits (6 and 9) leads to $94.
Comparing the results, Item C has the highest discounted price of $94.

Q: 27 A renowned alchemist prepares a magical potion with four ingredients: Moonflower (M), Nightshade (N), Dragon Scale (D), and Phoenix Tear (P). Each ingredient possesses unique properties, and the order in which they're added determines the potion's final effect. Here's what we know:

- Adding M first leads to an invigorating potion.
- Adding D first leads to a tranquillizing potion.
- Adding N first leads to a truth-telling potion.
- Adding P first leads to a transformative potion.

Furthermore, certain ingredient combinations result in specific outcomes:

- M + N = poisonous
- (invisible) + M = hallucinogenic
- P + N = invisibility
- P + M = flight

Given the following sequence of ingredient additions: P -> N -> M, what will be the final effect of the potion?

A. Invigorating and transformative
B. Truth-telling and poisonous
✓ Correct Ans C. Hallucinogenic
D. Impossible to determine with the given information

Explanation

Analyze the sequence step-by-step:

P added first: Sets the base effect as transformative.
N added second: Triggers the combination P + N, resulting in invisibility.
M added last: This ingredient affects the potion based on the existing state. Since the potion is currently invisible, adding M doesn't trigger its first effect (invigorating) but instead interacts with the existing invisibility:

M + (invisible) = hallucinogenic according to the given combinations.

Q: 28 Three alien species use unique currencies: Zargs (Z), Glurps (G), and Sprockets (S). Their exchange rates are represented by inequalities:

Z > G
G > S

Based on these inequalities, we can determine ...

A. 1 Z can be exchanged for 2 G.
B. 1 G can be purchased with 3 S.
C. You can always get more S by exchanging Z.
✓ Correct Ans D. More information is needed to determine the specific exchange rates.

Explanation

The inequalities only tell us the relative values of the currencies: Z is worth more than G, and G is worth more than S. However, they don't provide any specific conversion rates or ratios between the currencies.
While option (a) and (b) might seem plausible, there's no guarantee they're accurate based on the given information. Similarly, option (c) can't be confirmed, as we don't know the exact exchange rates for multiple S against one Z. Therefore, to determine the specific exchange rates and whether you can always get more S by exchanging Z, more information is needed. The inequalities offer a general direction about the currencies' values but lack the details to answer definitively.

Q: 29 You walk into a clothing store and find a beautiful jacket priced at $100. The store has a promotion: one item gets 50% off, another item gets 25% off, and any remaining items are full price. You also have a coupon for $10 off any purchase. How can you maximize your savings on the jacket?

    A. Buy the jacket only and use the coupon.

✓ Correct Ans   B. Buy the jacket along with a cheap item and apply the 50% discount to the jacket.

    C. Buy the jacket along with another item and apply the 25% discount to the jacket.

    D. The savings are equal regardless of your purchase.

Explanation

Using the coupon only provides $10 off the $100 jacket, total $90.
Buying the jacket with a cheap item (e.g., $5 accessory) allows you to apply the 50% discount to the more expensive jacket.
This reduces the jacket price to $50 and you can still use the coupon on the entire purchase (jacket + accessory), bringing the total to $40 + $5 = $45.
Therefore, buying the jacket with a cheap item and applying the 50% discount maximizes your savings compared to other options.

Q: 30 Tiffany is about going out on a winter day. She is perplexed about how to dress. She has the choice of wearing a blouse or shirt; a skirt or a pair of jeans or plain pants.
Because the weather is so cold she has the choice of wearing a pullover, a sweater, a parka or a blazer. She can then decide whether to wear a head warmer or a scarf.
In how many combinations can Tiffany dress from 'head to-toes' when going out?

    A. 36

✓ Correct Ans   B. 48

    C. 24

    D. 12

Explanation

Gary has choices in 3 major areas:
Top attire: Blouse or shirt (2)
Bottom attire: Jeans, skirts, or pant trousers (3)
Covering: pullover, sweater, parka, or blazer (4))
Head: warmer, scarf (2)
So, all his choices are 2 X 3 X 4 X 2 = 48 choices.

Q: 31 Seats at a musical concert were arranged such that each successive row had 1 seat more than the one preceding it. In which row will 33 people be sitting if the first row has 16 people sitting?

- A. 19th
- B. 17th
- C. 15th
- ✓ Correct Ans D. 18th

Explanation

Each row increases by 1
The first row has 16 people seated
The row with 33 people is **a**
a + (16 -1) = 33
a + 15 = 33
a = 33 -15 = 18
Ans : 18th row.

Q: 32 In a scholarship test, Rob was placed ahead of Larry who scored 67%. Nim scored 60% but was not shortlisted for the interview. Out of four friends, Rob, Dan, Nim and Larry, only Rob and Dan, who were ahead of Nim were shortlisted for the interview.
With the information above, which one of the following is true?

- A. Nim was placed between Rob and Larry
- ✓ Correct Ans B. Dan scored more than 67%.
- C. Larry was placed between Rob and Dan
- D. Nim was placed third among the four friends

Explanation

Rob and Dan were ahead of Larry. Larry got 67% but didn't get shortlisted. So Dan got more than 67% as Dan was shortlisted.

Q: 33 Audrey is faster than Hope, and Edith is faster than Audrey. This trio loves watching movies when together. It is Hope who often chooses the movie they will watch.
If all the above are facts, which of the following statements is also true?

- A. Hope is the fastest among the three individuals
- B. Audrey is the fastest among the three people
- ✓ Correct Ans C. Before watching a movie together, Hope often selects it
- D. Hope hates to watch a movie she hasn't chosen

Explanation

Since Edith is faster than Audrey, who is faster than Hope, statements A and B are wrong. The last statement supports C, whereas one can't tell whether D is correct given the above facts.

Q: 34 If their memories serve them right, Ann, Gideon, and Angeline agree that their friend Eric resides in estate D. They also agree that his apartment is on the 4th floor. Two of them agree that the number after the hyphen is 3, whereas the one that follows is 5. Two of them also agree that the last number is 8.

If the apartment number comprises the estate name, floor number, a hyphen, and 4 numbers, which is the most probable number for Eric's apartment ?

- A. D4-3587
- B. D4-5218
- C. D4-4358
- ✓ Correct Ans D. D4-3568

Explanation

A is wrong since the last number is 8. B is wrong since the number after the hyphen is most likely 3, which also nullifies C. Other than the unknown figure, the ones in D meet all the stated criteria hence most likely the apartment number.

Q: 35 Despite being a great event planner, Daniel has noticed that locals don't use his services often. It turns out that they think that Daniel won't accept gigs if the event is small.

If all the above are facts, which of the following statements can make him change this perspective?

- A. Relocating his offices and trying to woe a different clientele
- ✓ Correct Ans B. Using local channels such as the local magazine to advertise his business
- C. Advertising his business during huge global exhibitions and workshops
- D. Only offers discounts and coupons to his big clients

Explanation

Since the appropriate strategy should revolve around making the locals feel that even people hosting small events are eligible for his planning services, B will do the trick.

Q: 36   Find out that three figures from the given figures by which an equilateral triangle can be made.

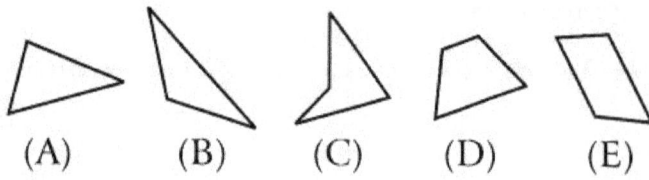

        A. BCD
        B. ACD
✓ Correct Ans C. CDE
        D. BDE

Explanation

Q: 37   Replace question mark (?) with appropriate number from amongst the options.
48, 24, 42, 21, 36, 18, 30, ?, ?

✓ Correct Ans A. 15, 24
        B. 12, 21
        C. 18, 26
        D. 21, 25

Explanation

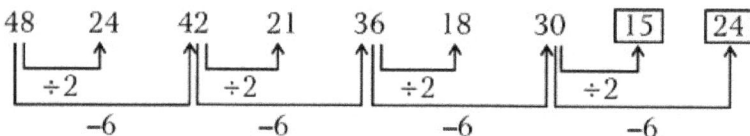

Q: 38 Figure 1 shows a rectangular piece of paper.
The ratio of its length to its breadth is 4 : 3.
In Figure 2, the piece of paper is folded and cut along the dotted line.
Figure 3 shows the cut-out, C, and the remaining area of paper, R, which is square.

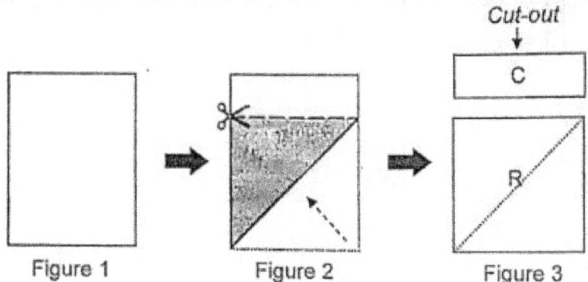

Figure 1   Figure 2   Figure 3

What is the ratio of the length to the breadth of C?

A. 4:9

B. 2:1

✓ Correct Ans C. 3:1

D. 4:1

Explanation

3:1

Q: 39 Use the rule of the pattern and find the missing number which will replace the question mark.

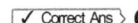

A. 21

✓ Correct Ans B. 25

C. 45

D. 36

Explanation

(B): 4 x 9 = 36 → √36 = 6
9 x 16 = 144 → √144 = 12
Similarly, 16 x **X** → √16**X** = 20
On squaring :
16**X** = 400 $\therefore x = \dfrac{400}{16} = 25$

Q: 40   Find the missing figure from the given alternatives.

A.

B.

C.

✓ Correct Ans D.

Explanation

(D): The triangle moves two steps in anticlockwise direction while the circle moves one step in anticlockwise direction.

Preptive Prepare Better
Series J Reading Comprehension Trial Test

Answers and Solutions

Q: 1 How do flamingoes construct their nests, according to Extract A?

- A. Using twigs and branches
- B. Digging holes in the ground
- ✓ Correct Ans C. Making intricate nests of mud and vegetation
- D. Repurposing abandoned bird nests

Explanation

Extract A states that flamingoes construct intricate nests made of mud and vegetation, often in shallow waters to protect their eggs from predators.

Q: 2 How do conservation efforts aim to protect flamingo populations, according to Extract B?

- A. By promoting habitat destruction
- B. By increasing pollution levels
- ✓ Correct Ans C. Through establishing protected areas and regulating pollution
- D. By ignoring human impact on flamingo habitats

Explanation

Extract B states that conservation efforts aim to protect flamingo populations by establishing protected areas and regulating pollution to mitigate human impact on their habitats.

Q: 3 What demonstrates the resilience of flamingoes, as discussed in Extract B?

- A. Their ability to migrate long distances
- B. Their reliance on human intervention
- ✓ Correct Ans C. Their adaptation to extreme temperatures and harsh conditions
- D. Their dependency on artificial habitats

Explanation

Extract B highlights the resilience of flamingoes by mentioning their adaptation to withstand extreme temperatures and harsh environmental conditions.

Q: 4 According to Extract A, what is the significance of the vibrant pink plumage of flamingoes?

- A. It serves as a defence mechanism against predators
- B. It provides camouflage in their natural habitats
- ✓ Correct Ans C. It derives its color from the pigments in their food and serves as a distinctive feature
- D. It indicates their age and maturity level

Explanation

Extract A mentions that the vibrant pink plumage of flamingoes derives its color from the pigments in their food, making it a distinctive feature of these birds.

Q: 5 How do flamingoes utilize their uniquely shaped bills, as discussed in both extracts?

- A. For territorial displays
- B. For vocalizations and mating calls
- ✓ Correct Ans C. For foraging and filter-feeding
- D. For building nests and shelter

Explanation

Both extracts highlight that flamingoes use their uniquely shaped bills for foraging and filter-feeding, enabling them to extract nutrients efficiently from shallow waters.

Q: 6 What culinary landscape is characterized by its reliance on locally sourced ingredients, as described in Extract A?

- A. North African cuisine
- B. West African cuisine
- C. East African cuisine
- ✓ Correct Ans D. Central African cuisine

Explanation

Extract A discusses how Central African cuisine relies on locally sourced ingredients such as root vegetables, leafy greens, and freshwater fish, contributing to its simple yet flavorful cooking techniques.

Q: 7 According to Extract A, what is the overarching theme of African cuisine?

- A. Fusion of European and Asian influences
- B. Celebration of seasonal ingredients
- C. Emphasis on bold and spicy flavors
- ✓ Correct Ans D. Diversity, flavor, and tradition

Explanation

Extract A concludes that African cuisine is a celebration of diversity, flavor, and tradition, encapsulating the rich culinary heritage of the continent.

Q: 8 According to Extract B, what contributes to the bold and spicy flavors of African dishes?

- A. Use of bland ingredients
- B. Incorporation of fresh herbs
- ✓ Correct Ans C. Skillful use of aromatic spices and chili peppers
- D. Absence of seasoning

Explanation

Extract B discusses how the bold and spicy flavors of African dishes are achieved through the skilful use of aromatic spices, chili peppers, and herbs, enhancing the overall sensory experience of the cuisine.

Q: 9 What distinguishes the culinary traditions of North Africa, as described in Extract A?

- A. Emphasis on bold and spicy flavors
- B. Reliance on locally sourced ingredients
- ✓ Correct Ans C. Influence of Mediterranean and Arab cultures
- D. Preference for fusion cuisine

Explanation

Extract A highlights that North African cuisine is shaped by its historical interactions with various cultures, including Arab, Berber, and Mediterranean influences, contributing to its unique flavor profile.

Q: 10 What role does communal dining play in the context of African cuisine, according to Extract B?

- A. It encourages individual dining experiences
- B. It limits the sharing of food among family and friends
- ✓ Correct Ans C. It fosters a sense of community, hospitality, and togetherness
- D. It discourages celebrations and festivals

Explanation

Extract B highlights how communal dining in African cuisine fosters a sense of community, hospitality, and togetherness, with meals often shared among family and friends during celebrations and festivals.

Q: 11 In which extract do animals navigate the seas in harmonious pods, showcasing social intricacies and a level of intelligence akin to human societal complexities?

- A. Extract A
- ✓ Correct Ans B. Extract B
- C. Extract C
- D. Extract D

Explanation

The correct answer is (b) Extract B. It discusses dolphins as captivating ambassadors of the sea, highlighting their playful demeanor, agility, and social intricacies comparable to human societal complexities.

Q: 12 Which extract features animals revered across diverse cultures as symbols of foresight and intuition, with an ability to reveal what remains hidden to others?

- A. Extract A
- B. Extract B
- ✓ Correct Ans C. Extract C
- D. Extract D

Explanation

The correct answer is (c) Extract C. It talks about owls as enigmatic creatures of the night, embodying an aura of mystique and wisdom, and being revered as symbols of foresight and intuition.

Q: 13 In which extract do animals undergo a breathtaking metamorphosis, transforming from unassuming caterpillars to resplendent creatures with vibrant hues and intricate patterns?

- A. Extract A
- B. Extract B
- C. Extract C
- ✓ Correct Ans D. Extract D

Explanation

The correct answer is (d) Extract D. It discusses butterflies as nature's ephemeral marvels that undergo a miraculous metamorphosis, transforming from caterpillars to resplendent creatures with vibrant hues.

Q: 14 Which extract describes animals that flutter gracefully through gardens and meadows, scattering joy with their ephemeral flights and leaving traces of exquisite existence?

- A. Extract A
- B. Extract B
- C. Extract C
- ✓ Correct Ans D. Extract D

Explanation

The correct answer is (d) Extract D. It talks about butterflies fluttering through gardens and meadows, leaving traces of joy with their ephemeral flights.

Q: 15 In which extract do animals symbolize the cyclical rhythms of life and serve as timeless reminders to embrace change and revel in fleeting moments of sublime magnificence?

- A. Extract A
- B. Extract B
- C. Extract C
- ✓ Correct Ans D. Extract D

Explanation

The correct answer is (d) Extract D. It discusses butterflies as nature's ephemeral marvels symbolizing cyclical rhythms of life and reminding us to embrace change and revel in fleeting moments of sublime magnificence.

Q: 16 What does the phrase "Be Thou my vision" signify in the poem?

- ✓ Correct Ans A. Seeking clarity of sight
- B. Yearning for material success
- C. A wish for physical strength
- D. Longing for artistic inspiration

Explanation

The repetition of "Be Thou my vision" suggests the speaker's plea for spiritual guidance and clarity in perceiving the world around them.

Q: 17 What does the speaker imply by calling the Lord "Thou my great Father, I Thy true son"?

- A. The speaker seeks protection from their parents
- B. The speaker desires material inheritance
- ✓ Correct Ans > C. The speaker acknowledges a spiritual relationship
- D. The speaker wishes for earthly comforts

Explanation

The line refers to a deeper spiritual connection between the speaker and the divine, likening it to a familial bond.

Q: 18 What sentiment does the line "Thou my soul's shelter, Thou my high tower" convey?

- ✓ Correct Ans > A. Seeking refuge and protection in the divine
- B. A desire for physical strength
- C. The speaker's admiration for nature
- D. Longing for financial stability

Explanation

The line expresses the speaker's plea for the Lord to be their shelter and refuge, emphasizing the need for divine protection.

Q: 19 How does the speaker view worldly possessions and praise from others?

- A. They are considered valuable and sought after
- ✓ Correct Ans > B. They are disregarded as unimportant
- C. They are seen as essential for happiness
- D. They are sought after but with caution

Explanation

The speaker expresses a lack of interest in material riches and praises from others, emphasizing the importance of spiritual wealth.

Q: 20 What emotion does the concluding line "Still be my vision, O Ruler of all" evoke?

- A. Fear
- ✓ Correct Ans > B. Serenity
- C. Sadness
- D. Confusion

Explanation

The final line conveys a sense of peace and tranquility, expressing the speaker's desire for the divine presence to remain their guiding vision.

Q: 21 Which one fits in (1) ?

- A. Sentence A
- B. Sentence B
- ✓ Correct Ans **C. Sentence G**
- D. Sentence H

Explanation

The next parts of the paragraph says about the urban planning, engineering skills etc. therefore, this sentence should be a statement proving hat fact. There fore correct sentence is **"The city of Harappa, for which…"**

Q: 22 Which one fits in (2) ?

- ✓ Correct Ans **A. Sentence A**
- B. Sentence B
- C. Sentence C
- D. Sentence D

Explanation

The correct sentence should be an introduction to the advanced features of the city as a guideline for the next sentence. So, the sentence should be "The Harappa civilization was characterized …"

Q: 23 Which one fits in (3) ?

- A. Sentence A
- B. Sentence D
- ✓ Correct Ans **C. Sentence B**
- D. Sentence E

Explanation

The sentence after explains about the ornaments. Thus, the correct sentence should be related to ornaments. Thus, the correct answer is **"These seals, made of steatite…"**

Q: 24 Which one fits in (4) ?

- A. Sentence A
- B. Sentence B
- C. Sentence H
- ✓ Correct Ans **D. Sentence E**

Explanation

The next parts of the paragraph directly has the word "Indus script". Thus, the correct sentence is **"One of the most intriguing aspects…."**

Q: 25 Which one fits in (5) ?

- A. Sentence A
- B. Sentence B
- ✓ Correct Ans C. Sentence C
- D. Sentence D

Explanation

The paragraph describes about the downfall of the civilization. Thus, the correct answer is **"The decline of the Harappa civilization…."**

Q: 26 Which one fits in (6) ?

- A. Sentence A
- B. Sentence B
- C. Sentence C
- ✓ Correct Ans D. Sentence D

Explanation

The correct answer should be aligned with the words of the sentence before, "archaeological legacy". Therefore the correct answer is **"The well-planned cities, advanced engineering…."**

Q: 27 Which one fits in (7) ?

- A. Sentence A
- B. Sentence G
- ✓ Correct Ans C. Sentence H
- D. Sentence C

Explanation

The end sentence should be a overall conclusion about the civilization. So. The correct answer is, "It leaves behind…"

Q: 28 What is the significance of the line "the prize we sought is won"?

- ✓ Correct Ans A. Goal of the journey has been achieved
- B. The journey has ended
- C. The journey is tiresome
- D. None of the above

Explanation

The line suggests that the goal or objective of the journey has been achieved.

Q: 29 Why does the speaker refer to the ship as "the vessel grim and daring"?

✓ Correct Ans **A.** Ship's resilience and courage
B. The weaknesses of the ship
C. The long journey of the ship
D. None of the above

Explanation

The speaker emphasizes the ship's resilience and courage in the face of challenges.

Q: 30 Why does the speaker call the captain "dear father"?

A. They are strangers
✓ Correct Ans **B.** To convey an emotional connection
C. They have known each other
D. None of the above

Explanation

The speaker addresses the captain as "dear father" to convey a personal and emotional connection.

Preptive Prepare Better
Series J Mathematical Reasoning Trial Test

Answers and Solutions

Q: 1 The total cost of 5 shirts and 6 pants is $210.55.
The total cost of 3 shirts and 2 pants is $91.45.
All the shirts are identical and all the pants are identical.

Find the cost of 1 shirt.

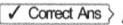 **A.** $15.95

B. $63.80

C. $15.90

D. $274.35

Explanation

5S + 6P = 210.55 ... (1)
3S + 2P = 91.45 ... (2)

(2) x 3 - (1) :

9S + 6P = 91.45 x 3 = 274.35
5S + 6P = 210.55

4S = 274.35 - 210.55 = 63.80
S = 15.95

Q: 2

Sam had some apples. He sold 210 apples in the afternoon and $\frac{3}{7}$ of the remaining apples in the evening. In the end, he had $\frac{1}{3}$ of the total number of apples left.

How many apples did he have at first?

    A. 505 apples

✓ Correct Ans   B. 504 apples

    C. 210 apples

    D. 42 apples

Explanation

Sam had 'n' apples at start.
He sold in afternoon = 210
He sold in evening = $3/7$ (n - 210) = $3/7$ n - 90

Left = n - 210 - ($3/7$ n - 90)
= n - 210 - $3/7$ n + 90 = $4/7$ n - 120

$4/7$ n - 120 = $1/3$ n
$4/7$ n - $1/3$ n = 120
$12/21$ n - $7/21$ n = 120
$5/21$ n = 120
n = 120 x ($21/5$) = 24x21 = 504

Q: 3

The table below shows the number of girls and boys who wear and do not wear spectacles in Class 6W.

	Number of girls	Number of boys
Wear spectacles	8	?
Do not wear spectacles	7	11

The **ratio** of the number of girls to the number of boys in 6W is 3 : 5.

Find the number of boys who wear spectacles in 6W.

    A. 5 boys

    B. 25 boys

✓ Correct Ans   C. 14 boys

    D. 15 boys

Explanation

Let us assume there are 'n' boys who wear spectacles.
Total boys = n + 11
Total girls = 15
Girls : Boys = 3:5. This means there are 3 parts girls and 5 parts boys.
3 parts = 15; 1 part = 5; 5 parts = 25
n + 11 = 25; n = 14

Q: 4 The price of 1 Dino melon was $20. A shopkeeper gave 1 Dino melon free for every 4 Dino melon that were bought. Lisa spent a total of $620 on the Dino melons.
How many Dino melons did Lisa get?

✓ Correct Ans **A.** 38
B. 31
C. 7
D. 20

Explanation

Let us use grouping concept.
1 group of 5 melons (including one free) cost $80.

In $620, there are 7 groups = 35 melons
Remaining money = $620 - 7 x $80 = $60.
3 additional melons can be bought by $60.
Total melons = 35 + 3 = 38

Q: 5 At a carnival, the children were put into two groups. $\frac{2}{3}$ of Group A were boys and $\frac{3}{5}$ of Group B were girls.
Group B had twice as many children as Group A.
There were 26 fewer boys in Group A than Group B.

What fraction of the total number of children at the carnival were boys? Give your answer in the simplest form.

A. 2/5
B. 5/6
✓ Correct Ans **C.** 22/45
D. 8/45

Explanation

Group A = n; Group B = 2n; Total = 3n
Boys in A = 2/3 n
Boys in B = 2/5 of 2n = 4/5 n
Total Boys = (2/3 + 4/5) n = 22/15 n

22/15 is what fraction of 3 ?
22/15 divided by 3 = 22/15 x 1/3 = 22/45

Q: 6 Janet spent 30% of her money on 7 cupcakes and 4 cookies on Monday.
The cost of each cupcake was twice the cost of each cookie.
She bought some more cupcakes and another 10 cookies with $6/7$ of her remaining money on Tuesday.
How many cupcakes did she buy on Tuesday ?

    A. 26

✓ Correct Ans   B. 13

    C. 25

    D. 28

Explanation

1 cupcake = 2 cookies
30% money spent on 18 cookies
Remaining money = 70%
Tuesday spent = 60%
Janet can afford 36 cookies.
Already bought = 10 cookies
Can buy more = 26 cookies = 13 cupcakes
She bought 13 cupcakes on Tuesday.

Q: 7 Adrian and Bert had a total of $4563. After Adrian spent $\frac{1}{4}$ of his money and Bert spent $\frac{2}{3}$ of his money, they had an equal amount of money left.
How much money was Bert left with?

✓ Correct Ans   A. $1053

    B. $1035

    C. $1005

    D. $1030

Explanation

A + B = 4563

$3/4$ A = $1/3$ B

A = $4/9$ B
$13/9$ B = 4563
B = 3159

Bert is left with $1/3$ B = 1053

Q: 8   The picture shows the duration of rain in a region.

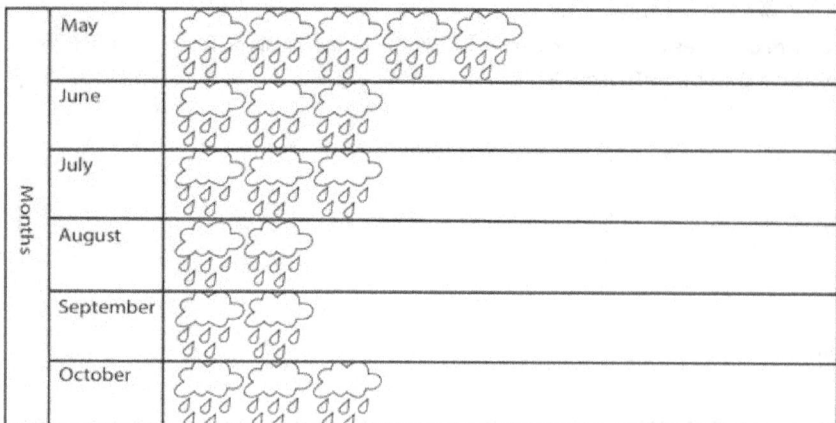

 = 50 minutes

Which of the following is a correct statement?

    A. There are 100 more minutes of rain in August than in September.

    B. There are 150 more minutes of rain in October than in May.

    C. There are 100 fewer minutes of rain in May than in June

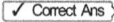 **D.** There are 50 more minutes of rain in June than in August.

Explanation

A. There are 100 more minutes of rain in August than in September. Wrong statement.
B. There are 150 more minutes of rain in October than in May. Wrong statement.
C. There are 100 fewer minutes of rain in May than in June. Wrong statement.
D. There are 50 more minutes of rain in June than in August. Correct statement.

Q: 9 Jack has $y for pocket money. Krishan has thrice as much pocket money as Jack. Latiff has $10 less than Krishan.

What is the total amount of pocket money the three boys have in terms of y?

✓ Correct Ans **A.** $(7y-10)
B. $(7y-20)
C. $(8y-6)
D. $(9y-9)

Explanation

Amt Jack has = $y
Amt Krishan has = $3y
Amt Latiff has = $(3y - 10)
Total Amt = $(y+3y+3y - 10) = $(7y - 10)

Q: 10 Adam has three fewer one-dollar coins than ten-cent coins. The total value of the ten-cent coins is $7.80 less than the total value of the one-dollar coins.
How many coins does Adam have in all?

A. 22
B. 20
✓ Correct Ans **C.** 21
D. 24

Explanation

	Count	Value
10 c coin	n	0.1 x n
$1 coin	n - 3	n - 3

n - 3 - 0.1n = 7.80
0.9 n = 10.80; n = 12

Total coins = n + n - 3 = 21

Q: 11 In a library $\frac{3}{5}$ of the books were English books and $\frac{3}{8}$ of the remaining books were Chinese books. The rest were Malay books. There were 133 more English books than Malay books in the library. How many books were there in the library altogether?

✓ Correct Ans A. 380
B. 228
C. 308
D. 448

Explanation

Eng = $3/5$
Remaining = $2/5$
Chinese = $2/5 \times 3/8 = 3/20$
Malay = $1 - 3/5 - 3/20 = (20 - 12 - 3) / 20 = 1/4$
Eng - Malay = $3/5 - 1/4 = 7/20$
$7/20$ is 133. $1/20$ is 19. Whole is $19 \times 20 = 380$

Q: 12 The angles of a quadrilateral are $(p+25)°$, $2p°$, $(2p-15)°$ and $(p+20)°$. What is the value of the largest angle?

A. 105°
✓ Correct Ans B. 110°
C. 115°
D. 135°

Explanation

The sum of angles in a quadrilateral is 360. So, $6p + 30 = 360$
$p = 330/6 = 55$
The biggest angle is $2p = 110°$

Q: 13 What are the correct coordinates of the vertices of the given polygon?

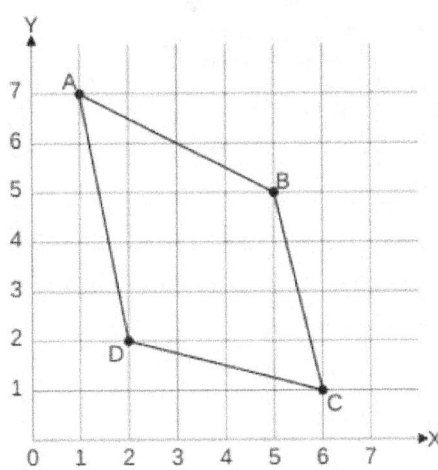

- A. A(1, 4) B(5, 5) C(6, 1) D(2, 2)
- B. A(1, 7) B(5, 5) C(6, 1) D(3, 2)
- ✓ Correct Ans C. A(1, 7) B(5, 5) C(6, 1) D(2, 2)
- D. A(1, 7) B(5, 4) C(6, 1) D(2, 2)

Explanation

X-axis then Y-axis.

Q: 14 Which of the following statements is correct?

- ✓ Correct Ans A. $(2^3)^2$ and $(3^2)4^4$ are not the same
- B. $(2^3)^2$ and $(3^4)^2$ are the same
- C. $(6^{49})^2 = 6^{492}$
- D. $(8^7)^3 = 8^{10}$

Explanation

A. $(2^3)^2$ and $(3^2)4^4$ are not the same : correct statement
B. $(2^3)^2$ and $(3^4)^2$ are the same : wrong
C. $(6^{49})^2 = 6^{492}$: wrong
D. $(8^7)^3 = 8^{10}$: wrong

Q: 15 The librarian spent $1888 buying books for the school library during the promotion as shown below. How many books did she buy altogether?

Books
At $16 each

PROMOTION
Buy any 18 books
and
get the 19th and 20th book at
half price each

✓ Correct Ans A. 124
B. 146
C. 144
D. 68

Explanation

Use grouping concept.

Pay for 19 books, get 20.
1 group has 20 books, costs 16x19 = $304
$304 fetches 20 books.
$1888 fetches = 6x20 books + $64 remaining
With remaining $64, we can get 4 more.
Total books = 120+4 = 124

Q: 16 Consider the following 3 statements.

1. Price of 4 toffees is $2
2. Price of 1 chocolate and 2 lollipops is $3.4
3. When we subtract price of two lollipops from the price of a toffee, the remaining price is equal to ½ price of a lollipop.

Calculate the total amount you have to pay if you buy 1 toffee, 1 chocolate and a lollipop.

A. $2.10

✓ Correct Ans B. $3.70

C. $4.50

D. $5.20

Explanation

Toffee=T, Chocolate=C, Lollipop=L
By statement 1 : 4T = $2
T= $2/4 = $0.5 (Price of a toffee)
By statement 3 :
T-2L=1/2 L
Multiply above equation by 2 .
2T-4L= L
2T=5L
As T= $0.5, 2 ×$0.5=5 L
$1=5L
L=$0.2
By statement 2 :
C+2L=$3.4
As L= $0.2, C+2 ×$0.2= $3.4; C=$3
Total=$0.5+$0.2+$3=$3.7

?

Q: 17

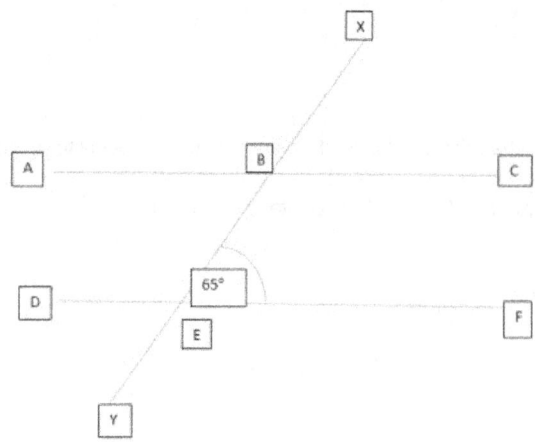

AC and DF are parallel lines. What is the value of ABX angle?

- A. 65°
- B. 115° ✓ Correct Ans
- C. 90°
- D. 105°

Explanation

BEF angle = 65°
ABE and BEF are alternate angles. So ABE angle = 65°.
XBY is a straight line. Sum of the angles on a straight line is equal to 180°.
So : ABE + ABX = 180°
65° + ABX = 180°
ABX = 115°.

Q: 18 In a farm there are 68 rabbits, 95 cattle and the number of hens is 25% of number of rabbits. If $1/34$ rabbits and 20% percent of cattle are transferred to another B farm, how many animals remain in Farm A?

- A. 160
- B. 159 ✓ Correct Ans
- C. 158
- D. 157

Explanation

Hens = 68 × $25/100$ = 17
Transferred rabbits = 68 × $1/34$ = 2
Transferred cattle = 95 × $20/100$ = 19
Remaining animals = (68+95+17)−(2+19) = 159

Q: 19 The picture shows an automatic fish feeder and its specification.

Specifications
- Container size: Moderate
- Dried Food and pellet maybe used
- A timer is used to arrange feeding time
- Use the latest technology to prevent food from getting moist or stuck in the container
- Can be operated manually or automatically
- Digital screen display

Automatic fish feeder

If Eng Wei decides to feed the fish 4 times a day in equal interval with the first feeding time at 7:35 a.m., at what time should he feed the fish for the third feeding?

A. Fishes are fed for the third time at 7:45 p.m.

✓ Correct Ans B. Fishes are fed for the third time at 7:35 p.m.

C. Fishes are fed for the third time at 7:25 p.m.

D. Fishes are fed for the third time at 7:15 p.m.

Explanation

1 day = 24 hours
Each feed = $24/4$ = 6 hours
Pattern: 6 hours
T1 = 7:35 a.m.
T2 = 7:35 a.m. + 6 hours = 1:35 p.m.
T3 = 1:35 p.m. + 6 hours = 7:35 p.m.
Hence, fish are fed for the third time at 7:35 p.m.

Q: 20 In the diagram below, M' is the image of M in an axis of reflection.

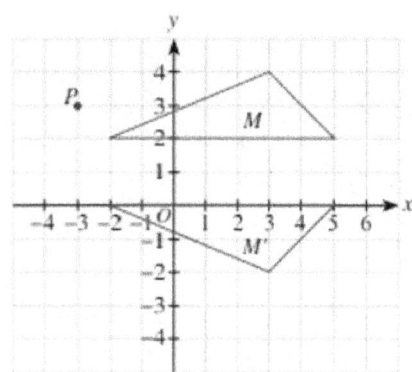

Determine the coordinates of P' under the same axis of reflection.

✓ Correct Ans **A.** (-3 , -1)
B. (3 , 1)
C. (-3 , 1)
D. (3 , -1)

Explanation

Axis of reflection y = 1,
Coordinates of P' are (–3, –1).

Q: 21 Mrs Lin prepared 160 chicken wings and some nuggets for a party. At one point during the party, an equal number of chicken wings and nuggets were eaten. 25% of the chicken wings and 20% of the nuggets were left. She then increased the number of chicken wings. After that, there was a total of 65 chicken wings.

How many nuggets did Mrs Lin prepare for the party?

A. 15
✓ Correct Ans **B.** 150
C. 120
D. 130

Explanation

At some point, 75% of cw and 80% of nuggets were eaten. They are same count.

80% of nugget = 75% of cw = 75% of 160 = 120
10% of nugget = 120/8 = 15
100% of nugget = 150

Q: 22

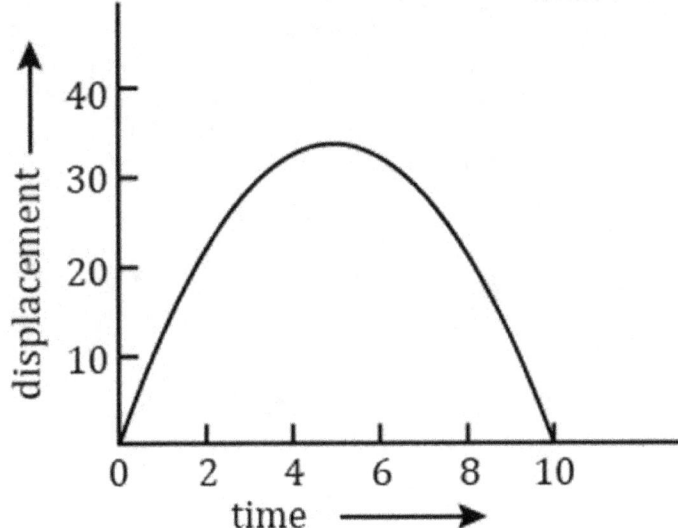

Layla throws a ball. The above graph represents the displacement- time graph of the ball (metre and sec). When it has a displacement of 20 m what is/are the times relevant for that displacement?

A. 2s

B. 4s

✓ Correct Ans C. 2s and 8s

D. 8s

Explanation

Find 20 m on the y axis. Move right and left until you intersect the graph. Now move downwards until you intersect the x axis. You intersect x axis at 2 points 2 s and 8 s.

Q: 23 Which of the following statements is/ are **incorrect**?

X. $4/5 + 1/5 + 2/5$ is more than 1
Y. $1/8 - 1/7$ gives a negative number
Z. $8/9$ is less than $6/7$

A. X and Y only
B. Y and Z only
C. Y only
✓ Correct Ans D. Z only

Explanation

X. $4/5 + 1/5 + 2/5 = 7/5$
X is correct.

Y. $1/8 - 1/7 = 7 - 8/56 = -1/56$
Y is correct.

Z. $8/9 = 56/63$; $6/7 = 54/63$
Z is incorrect.

Q: 24 Jessie had $80 less than Vincent at first. Vincent gave Jessie $60. The ratio of Jessie's money to Vincent's money now became 11 : 7.
How much did Vincent have at first?

A. $100
B. $110
✓ Correct Ans C. $130
D. $150

Explanation

Start : Vincent = n, Jessie = n-80
End : Vincent = n - 60, Jessie = n - 20

(n-20) : (n-60) = 11:7
(n-20)x7 = (n-60) x 11
7n - 140 = 11n - 660; 4n = 660 - 140 = 520; n = 130
Vincent had $130 at first.

Alternative method :
Vincent gave Jessie $60.
Difference became = $120.
Initially there was difference of $80.
At end, the difference is $120 - $80 = $40

Jessie = 11 parts, Vincent = 7 parts.
4 parts = $40
1 part = $10
Vincent at end = 7 parts = $70
Vincent gave Jessie = $60.
Vincent at first = $70 + $60 = $130

Q: 25   The figure on the right shows a 'star'. Find ∠x.

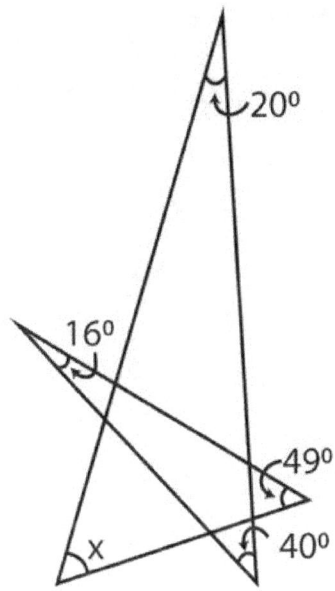

- A. 65°
- ✓ Correct Ans B. 55°
- C. 45°
- D. 75°

Explanation

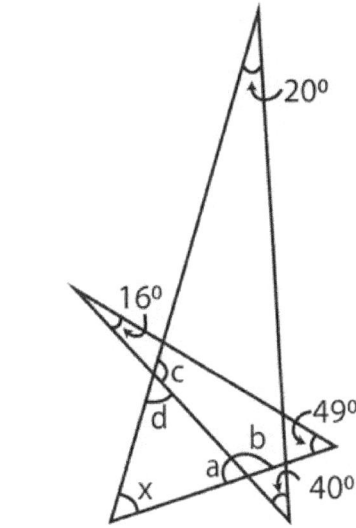

∠b = 180° − 16° − 49° = 115°
∠a = 180° − 115° = 65°
∠c = 180° − 20° − 40° = 120°
∠d = 180° − 120° = 60°
∠x = 180° − 65° − 60° = 55°

Q: 26 Jason had 40% fewer coins than Tom. Jason gave his sister some of his coins. As a result, the number of coins he had decreased by 25%.
How many coins did Jason have left if both had 58 coins altogether at the end?

✓ Correct Ans **A.** 18
B. 16
C. 15
D. 12

Explanation

Start : Tom = n, Jason = 0.6 n
End : Tom = n, Jason = 0.45 n
n + 0.45n = 1.45n = 58
n = 58/1.45 = 40
Jason end = 0.45n = 0.45x40 = 18

Q: 27 What is the perimeter of the given figure?

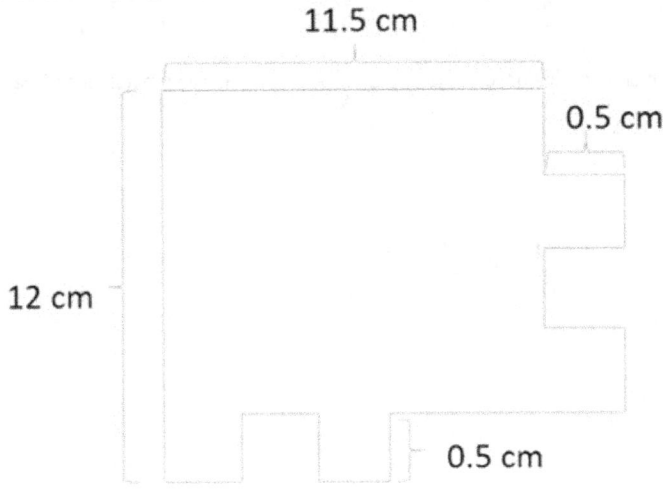

- A. 54 cm
- B. 52.5 cm
- ✓ Correct Ans C. 50 cm
- D. 48 cm

Explanation

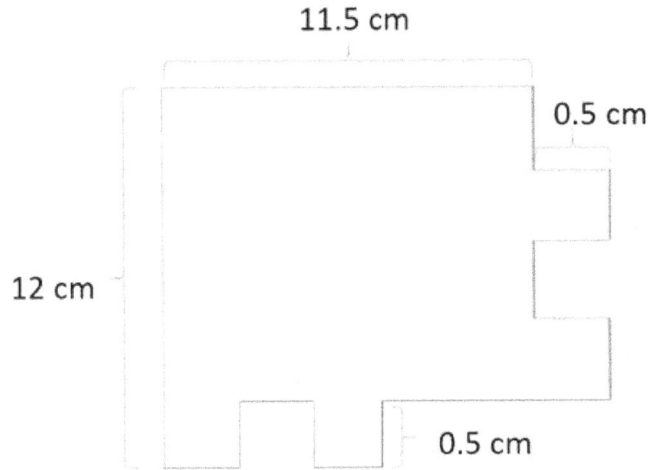

Highlighted horizontal lines must be added up to 11.5+0.5=12cm
Highlighted vertical lines must be added up to 12 cm
Perimeter = 12 ×4+0.5 ×4 = 48+2 = 50 cm

Q: 28 An ice cream cone costs $1.60 each. Each customer is entitled to buy **another** two at a discount of $0.30 off the original price each after buying three.

Owen has $24. Find the amount of money he will have in the end after buying the maximum number of ice cream cones.

- A. $0.50
- B. $0.30
- C. $0.40
- ✓ Correct Ans D. $0.20

Explanation

Group of 5 = 1.60x3 + 1.30 x 2 = $7.40
Group of 15 costs = 7.40x3 = $22.20
Left = $1.80
Buy one more at full price $1.60.
Left = 1.80 - 1.60 = 0.20

?

Q: 29 Roads A and B were of the same length. Each of the 21 street lamps on Road A was 10 m apart. Road B had 4 fewer street lamps than Road A. Find the distance between 2 street lamps on Road B.

- A. 14.5m
- B. 16.5 m
- C. 15.5 m
- ✓ Correct Ans D. 12.5 m

Explanation

21 – 1 = 20 (Total number of gaps on Road A)
20 × 10 m = 200 m (Length of both roads)
21 – 4 = 17 (Street lamps on Road B)
17 – 1 = 16 (Number of gaps on Road B)
200 m ÷ 16 = 12.5 (Distance between every 2 street lamps on Road B)
The distance between every 2 street lamps on Road B was 12.5 m

Q: 30 ABCD is a square piece of paper. A corner of the paper was folded to form triangle EFG. Find ∠y.

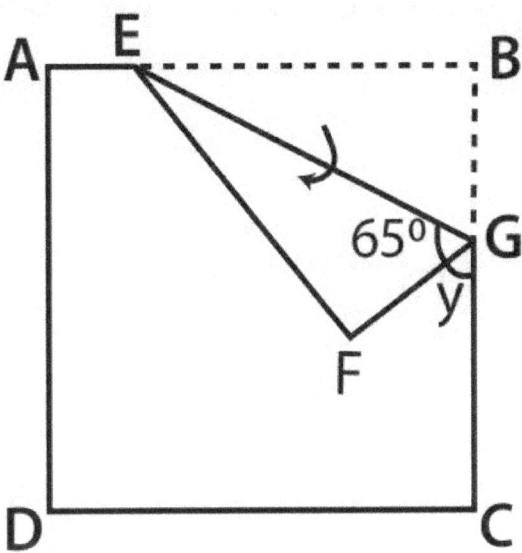

- **A.** 60
- **B.** 65
- ✓ Correct Ans **C.** 50
- **D.** 55

Explanation

∠EGF = ∠EGB (Triangle EFG = Triangle EBG)
∠y + ∠EGF + ∠EGB = 180°
180° − 65° − 65° = 50°
∠y = 50°

Q: 31 Lula is selling cakes to raise money for charity.
The graph shows how the amount of money left to raise is related to the number of cakes that Lula has sold.

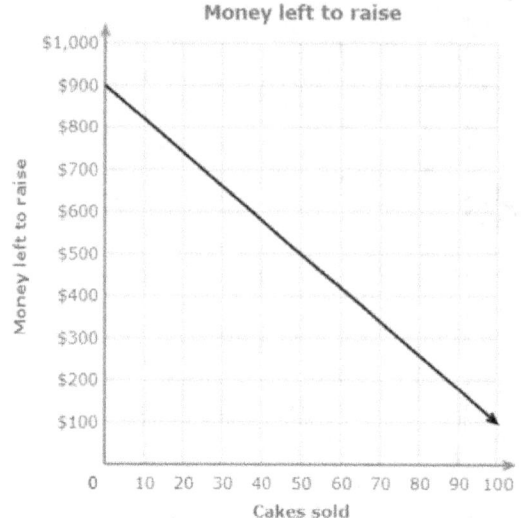

Guess the number of cakes Lula needs to sell in order to have just $500 left to raise.

A. 48
B. 37
✓ Correct Ans C. 50
D. 30

Explanation

Intersect the x-axis at 50 cakes. In order to have $500 left to raise, Lula needs to sell 50 cakes.

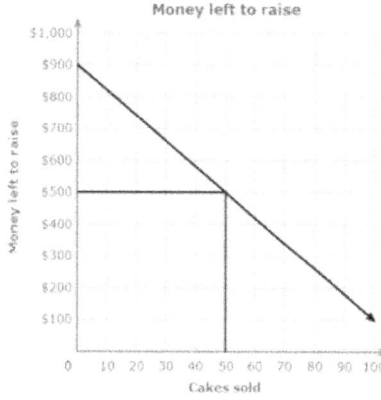

Q: 32 Which sign makes the statement true?

$$\frac{8}{9} \; ? \; \frac{4}{9} + \frac{1}{3}$$

✓ Correct Ans A. >

B. <

C. =

D. -

Explanation

Simplify the right side, the least common denominator is 9.

$$\frac{8}{9} \; ? \; \frac{4}{9} + \frac{1}{3}$$

$$\frac{8}{9} \; ? \; \frac{4}{9} + \frac{3}{9}$$

$$\frac{8}{9} \; ? \; \frac{7}{9}$$

$$\frac{8}{9} > \frac{7}{9}$$

The > sign makes the statement true.

Q: 33 Alicia read 24 pages of a story book every day while Brenda read 16 pages of the same story book every day. Brenda started reading the book on Monday, 2 days ahead of Alicia. On which day would both of them be on the same page?

A. Wednesday

B. Thursday

C. Friday

✓ Correct Ans D. Saturday

Explanation

	Mon	Tue	Wed	Thu	Fri	Sat
Brenda	16	32	48	64	80	96
Alicia			24	48	72	96

Q: 34

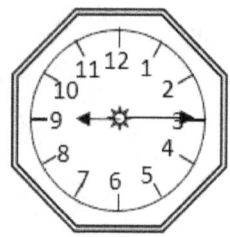

Henry wakes up and looks at the clock. He has 20 minutes to get ready for school. However, he got ready in 10 minutes and rode his bike to the school. It took 15 minutes to reach the school gates and 5 minutes to reach the class.
At what time does Henry arrive at the class?

✓ Correct Ans **A.** 9:45 a.m.

B. 9:50 a.m.

C. 9:35 a.m.

D. 9:55 a.m.

Explanation

9:15 + 10 minutes + 15 minutes + 5 minutes = 9:45 a.m.

Q: 35 Diagram below is an incomplete bill that shows the items bought by Alice.

Quantity	Item	Price per kilogram	Price
1 kg	Grape	$17	
4 kg	Guava		
		Total	$35

Alice went back to the store and bought 7kg more guavas. Calculate the price for all the guavas that Alice bought.

A. $50

B. $55.50

✓ Correct Ans **C.** $49.50

D. $49

Explanation

Find price of 1 kg of guava = total - price of grape = 35 – 17 = 18
18 ÷ 4 = 4.50 for 1kg of guavas
4 + 7 = 11kg of guavas that Alice bought
11x4.50 = $49.50

Preptive Prepare Better
Series J Thinking Skill Trial Test

Answers and Solutions

Q: 1 Tom and Harry are shooting marbles. The winner of each round collects a third of the marbles of the loser. Both started the game with 30 and 45 marbles respectively.

If Tom won the first round and Harry the second round, how many marbles would Tom have at the beginning of the third round?

✓ Correct Ans **A.** 30

B. 42

C. 45

D. 50

Explanation

At the start:
Tom = 30 marbles, Harry = 45 marbles.
If Tom won the first game, he gets $1/3$ of 45 marbles = 15 marbles from Harry
The second game start with Tom = 30 + 15 marbles = 45 marbles,
Harry = 45 − 15 marbles = 30 marbles
Harry won the second game so get $1/3$ of 45 marbles- 15 from Tom
So the third game starts with Tom having 45 − 15 marbles = 30 marbles and Harry has (30 +15) marbles = 45

Q: 2 A bookstore has 100 fiction novels and 50 non-fiction novels in stock. Each of the fiction novels sells for a price between $5 and $8, and each of the non-fiction novels sells for a price between $10 and $14.

Is the average price of all the novels in stock at the bookstore lower than $6?

A. Yes, lower than $6

✓ Correct Ans **B.** No, higher than $6

C. Same as $6

D. Can not be determined

Explanation

There are no specific prices for the novels. But there are twice as many fictions as non-fiction novels. Therefore the average price of the total novels will be closer to the price of the fiction than non-fiction.
To calculate the lowest possible total average, let's take the lowest price for all fiction and non-fiction: $5 and $10, the average will be
[($5 X 100) + $10 X50]/ 100 +50 = $1000 / 150 = $6.67

Q: 3 It is a boring day, Jerry is forced to stay indoors. So he decided to read. In his dad's library are some novels of different genres. These included 26 mysteries, 12 modern romances, 12 science fiction, 8 classical romances, and 10 general novels.

How many novels maximum can he read if half of the mysteries and a third of the science fiction that are his father's favourites are out of bound to him?

A. 47

B. 660

✓ Correct Ans C. 51

D. 48

Explanation

Available to Jerry are:

- Mysteries: 26 but half are out of bound, so he has 13
- Modern romances: 12
- Science fictions: 12: with a third out of bounds, so he has 8.
- Classical romances: 8, and
- General novels: 10.

So he has 13 + 12 + 48+ 8 + 10 = 51 novels

Q: 4 A Social Science Club is to attend a quiz competition with 3 representatives participating in the 3 subject areas. The club decided to hold a mock quiz among its members to pick the team to represent it at the competition.
The following are the results best performances of the mock quiz.

	Member	Scores in percentage (equal importance)		
		Politics	Philosophy	Economics
1	Rudolph	87	92	90
2	Betty	80	95	76
3	Pamela	91	90	90
4	Sheila	88	89	83
5	Juliet	70	83	85
6	Garry	60	91	82
7	Sam	92	97	96
8	Lucas	73	92	75

The club is to choose a 3-person team to represent it in the competition, based on the performance at the mock. Average score across subjects count.

Who are the best members of the team?

A. Sam, Pamela, Betty

B. Sam, Betty and Pamela

✓ Correct Ans C. Sam, Rudolph and Pamela

D. Sam, Betty and Rudolph

Explanation

As all are percentage scores, they can be added. We need to add and find top 3.

Q: 5   The opinion polls show that Mr McDonald of DDP is leading in the gubernatorial election poll by 45% to 35% over Mr Bailey of PCE. 20% of those polled are undecided.

Tim: "Mr McDonalds is set to be the next Governor."

Which one of the following sentences shows the mistake in Tim's statement ?

- A. Mr McDonald is leading his opponent in the opinion polls.
- ✓ Correct Ans B. The margin of undecided is large enough for Mr Bailey to possibly win the election
- C. Mr McDonald is more popular among those polled.
- D. Mr Bailey is 10 points behind Mr McDonald in the polls

Explanation

The 20% undecided can swing the results in Mr Bailey's favour.

Q: 6   Bill scored 86% on a Science test. And he performed better than Graham. Joe led the class in the test while Ted scored the least.
Given the information above, which one of the following **cannot be true**?

- A. Graham scored better than Ted
- B. Joe scored above 80 %
- ✓ Correct Ans C. Ted was placed between Bill and Graham
- D. Billy was placed between Joe and Graham

Explanation

Joe was the best-placed.
Ted is the least.
Bill scored 86% better than Graham.
So the placements are Joe. Bill, Graham and Ted.
Therefore Ted being placed between Bill and Graham is incorrect.

Q: 7 Four friends: Boyle, Ted, Henry and Dylan are members of a soccer team. They are goalkeeper, midfielder, striker, and defender but not in that order.
Two are left-footers. One of the left-footers is a goalkeeper, the other is a midfielder
Boyle, who is a striker, scores with his right foot as the dominant one
Ted is a left-footer and outfield player (meaning not goalkeeper).

Who is the other right footer?

✓ Correct Ans **A.** A defender

B. A winger

C. A Goalkeeper

D. A midfielder

Explanation

There are two left-footers: a goalkeeper and a midfielder
2 right footers: a striker and another person
The outfielders are: defenders, strikers and midfielders
Then Ted is a midfielder
The second right footer is a defender.

Q: 8 Shipping a cargo from one town to another involves taking some decisions. The consignment can be shipped either through the train or truck or by sea. The costs and capacity of the available transportation are as follows:

Means of Transport	Capacity	Cost of rental
Road	9kg	$28
Rail	15kg	$50
Sea	20kg	$60

If the weight of the consignment is 360kg, which means should be used?

A. Road

B. Rail

✓ Correct Ans **C.** Sea

D. All are same

Explanation

For the road @ 9kg he will need 360/9 loads = 40 X $28 = $1,120
Rail: @15kg, 360/15 loads= 24 X $50 = $1,200
Sea: @ 20kg= 360/20 loads= 18 X $60 = $1,080
Sea is most economical.

Q: 9 Statement : "If the floods don't end soon, most students will struggle to report to school due to inaccessibility. This is a big problem."

Jack: "Most students hail from areas affected by floods."
Lucy: "Accessibility to school is important to students."

Which of the following reasoning is correct?

   A. Jack Only
   B. Lucy Only
   ✓ Correct Ans C. Both Jack and Lucy
   D. Neither Jack nor Lucy

Explanation

Both statements are true since the fact states that most students will struggle to go to school if the situation continues.

Q: 10 The rules of a certain language are as follows.
- Any word that starts with a consonant has its first letter replaced by !
- A word that starts with a vowel has its first letter replaced by @.
- A word that starts and ends with a consonant has the first and the last letters replaced by # and $, respectively.
- A word that starts and ends with a vowel has the first and last replaced by % and ^, respectively.

Which of the following is a possible code for the word ARROW?

   ✓ Correct Ans A. @1234
   B. !1234
   C. #123$
   D. %123^

Explanation

Since the first letter is a vowel, one replaces it with @; hence, the probable one is A.

Q: 11 A new energy drink called "EnerGize" has been launched recently, claiming to boost energy levels by 200% in just 15 minutes. A group of university students decided to test its effectiveness. After consuming the drink, they felt more energetic during their study session. Based on this, Sarah concluded, "EnerGize can enhance anyone's energy levels by 200% within 15 minutes."

Which statement below best identifies a flaw in Sarah's conclusion?

✓ Correct Ans **A.** Sarah's conclusion is based on a small sample size of university students and may not represent the wider population's response to the energy drink.

B. EnerGize's claim about boosting energy by 200% is unrealistic and scientifically impossible.

C. Sarah's friends didn't feel any effects after consuming the energy drink, so it must not work for everyone.

D. Sarah's assumption is correct, as she and her friends experienced increased energy levels after consuming EnerGize.

Explanation

Sarah's conclusion is drawn solely from the experience of a small group of university students. This limited sample size may not accurately represent the broader population's response to the energy drink. Individual reactions can vary widely, and drawing sweeping conclusions from a small sample could be misleading.

Q: 12 If Scarlet is the eldest in her family and is now 25 years old, which of the following statements is also true if none of her siblings share the same birth year and the year of reference is 2022?

A. The third born hadn't been born in 2000

✓ Correct Ans **B.** No one among the siblings was born in 1995

C. The family has twins

D. The last born is at least 20 years old

Explanation

If it is 2022 and Scarlet is 25 years old, she was born in 1997. She is the firstborn; hence, B is correct. There aren't enough facts to support A. Since none of the siblings share a birth year, C is incorrect. Lastly, there isn't much information to support D.

Q: 13 A library organizes its books using a unique code system. Each code consists of a letter followed by three numbers. The numbers wrap around 999. The letter represents the genre (F for fiction, N for nonfiction, H for history), and the numbers represent the book's position on the shelf.
If the code "N521" is distantly followed by "N003" on the shelf, which of the following codes could be missing between them?

A. N499

B. H105

✓ Correct Ans **C.** N522

D. F520

Explanation

The missing code must be a nonfiction book (N) because it's between two nonfiction codes.
The missing code must have a higher number than 521 but lower than 003 (since numbers wrap around after 999).

Only option (c) N522 satisfies both conditions, maintaining the correct genre and numerical sequence.

Q: 14 A mysterious machine operates under these rules:

- It accepts words as input.
- It reverses the order of letters in the input word.
- If a vowel appears in the reversed word, it repeats that vowel at the end of the output.
- If no vowels appear in the reversed word, it adds "-tion" to the end of the output.

What would be the machine's output for the input "strange"?

A. egnarts

✓ Correct Ans B. egnartsae

C. egnart

D. egnartion

Explanation

Reverse the word: "egnarts"
Identify vowels: "a", "e"
Repeat the vowel: "egnartsae"

Q: 15 On a journey through the mystical land of Aethel, you encounter three enigmatic strangers: the Oracle, the Seer, and the Whisperer. You know:

- The Oracle always tells the truth.
- The Seer always lies.
- The Whisperer alternates between truth and falsehood with each statement.

You ask each stranger the same question: "Which of us is currently on the path to enlightenment?"

The Oracle says, "The Seer is enlightened."
The Seer says, "The Whisperer is not enlightened."
The Whisperer says, "The Oracle is lying."

Based on their pronouncements, who among the three may not be on the path to enlightenment?

✓ Correct Ans A. The Oracle

B. The Seer

C. The Whisperer

D. All three could be on the path.

Explanation

Oracle: Since the Oracle always tells the truth, the Seer must be enlightened (as stated by the Oracle).
Seer: As the Seer always lies, their statement about the Whisperer not being enlightened must be false, meaning the Whisperer is actually enlightened.
Whisperer: The Whisperer alternates truth and falsehood. However, there is no information if Oracle is on path to enlightenment.

Q: 16 Professor Elara, a renowned historian, investigates the reign of Queen Anya. She discovers:

- All of Queen Anya's loyal advisors were also skilled swordsmen.
- No skilled swordsman would ever betray their queen.
- Queen Anya's advisor Sir Gareth was not a skilled swordsman.

Can Professor Elara conclude with certainty that Sir Gareth betrayed Queen Anya?

- **A.** Yes, the conclusion is logically valid.
- **B.** Yes, but only with additional evidence.
- ✓ Correct Ans **C.** No, the conclusion is logically invalid.
- **D.** The conclusion is irrelevant to the investigation.

Explanation

The syllogism commits the fallacy of denying the converse. While it's true that all loyal advisors were swordsmen and no swordsman would betray the queen, the converse ("not loyal" implies "treacherous") does not necessarily hold true.

Q: 17 In the labyrinthine Library of Eldoria, each book contains a single true statement and a single false statement.

The librarian claims:
- If the book you are holding is true, then the book to the right is false.
- If the book you are holding is false, then the book to the left is true.

You pick up a book at random. Based on the librarian's claim, what can you definitely conclude about the book in your hand?

- **A.** It is true.
- **B.** It is false.
- ✓ Correct Ans **C.** It is either true or false, but we cannot know which.
- **D.** The statement about the librarian's claim is itself untrue.

Explanation

The librarian's statement creates a paradoxical situation. If the book in your hand is true, then the one to the right must be false, which could imply your book to be true again. If your book is false, then the one to the left must be true, but then the librarian's claim on your book might also be true, invalidating the initial assumption. Therefore, we cannot definitively determine the truth value of the book you hold based solely on the librarian's claim.

Q: 18 A gardening expert advises, "Rotating crops in your garden helps maintain soil fertility."

Which statement, if true, best supports the expert's advice?

- A. Some gardeners believe that using synthetic fertilizers eliminates the need for crop rotation.
- B. Limited space in urban gardens makes it challenging for individuals to rotate crops effectively.
- C. Modern irrigation systems can compensate for soil nutrient depletion, reducing the importance of crop rotation.
- ✓ Correct Ans D. Crop rotation prevents the depletion of specific nutrients in the soil, promoting long-term soil fertility.

Explanation

Crop rotation prevents the depletion of specific nutrients in the soil, contributing to long-term soil fertility, which supports the gardening expert's advice.

Q: 19 Since the winters will be extremely cold this year, Mike and Melody won't be going for their vacation in Spain. Instead, they plan to go to a new local restaurant. It means spending less than the initial budget. Unless they find availability, they won't spend the extra money in case they decide to go to Spain later.

If these are facts, which of the following statements is also true ?

- A. Mike and Melody have entirely given up on visiting Spain.
- B. It is now final that Mike and Melody will be going to the new local restaurant.
- ✓ Correct Ans C. The cold winters have affected Mike and Melody's decision to visit Spain.
- D. A trip to Spain is overpriced.

Explanation

The last statement nullifies option A. B is also wrong since it depends on availability. C is correct based on the first sentence. Lastly, whereas going to Spain is expensive, there aren't facts showing that it is overpriced.

Q: 20 Cosmos is an American sports club founded in 1976 by a sports enthusiast, Ferd Ackerman and his group of wealthy friends. They first formed a Baseball team, The Hard-hitters, in the group's home town, Detroit. Then, the group organized a Basketball team, the Jumpers.

The group first participated in American Football with the formation of a team in Cincinnati, the Cincinnati Kickers, in 1989. At that time the team was formed to participate in the sport's inner town league in the city. It was an innovation to inner town team sports and the new game of American Football. The group also formed the Jersey Dodgers to participate in the sport's Eastern Conference League. The San Francisco Smashers, its most successful team to date, was formed in 1992 to participate in the new sport's California State League. This team has gone on to win the State Championship 12 times since its formation.

After this club, came the formation of the Atlanta Cranes in 2008. After this, the group's appetite for new teams seems to have waned as no new club has been formed since.

Which sports team is the San Francisco Smashers?

    A. Basketball

    B. Baseball

✓ Correct Ans C. American football

    D. Squash

Explanation

The write-up started talking about the group's participation in American Football after the formation of the Cincinnati Kickers. The other accounts describe the group's participation in this sport (American Football).

Q: 21 This device should always be repaired by a professionally trained technician and handled carefully at all times.
Children must not be allowed to operate this device nor anybody under the influence of alcohol.
Repairs to this equipment may only be carried out by trained professionals or at the manufacturer's repair centres across the country.
Improper or unauthorized repairs can lead to considerable risks to users, including electrical shocks.

To which appliance would this user manual belong?

✓ Correct Ans A. A water heater

    B. A bookshelf

    C. A toy box

    D. An android tablet

Explanation

From the list of options, the appliance most susceptible to major electric shock waves is a water heater and the most complicated in repairs.

Q: 22 Every day Brian walks to school from his house, which is 1.4km away. However, on Saturdays, he goes to his father's workshop to help out before going for his football training in school; and then comes back home from school. His father's workshop is 700 metres from his home and 400 metres from his school.

How much is the difference between his walks to and from school on weekdays and Saturdays?

   A. 0.2km
   ✓ Correct Ans B. 0.3km
   C. 0.4km
   D. 0.56km

Explanation

On weekdays he walks 1.4km from home to school and 1.4 km from school to home.
Total = 2.8km
On Saturdays, he walks:

- Home to the workshop: 700m
- Workshop to school: 400m
- Total to get to school = 1.1km
- Then he walks 1.4km back home
- Total = 2.5km

So he walks more on weekdays:
2.8 – 2.5 km = 0.3km

Q: 23 The Grand Prix race at Silverstone started 25 minutes behind schedule because of a snow blizzard. The race was won by the Ferrari driver with a time of one hour and twelve minutes. The second-placed driver in the Mercedes car, came in 25 minutes after the Ferrari car crossed the line and the third-placed driver came in six minutes after the Mercedes car.
At what time did the third-placed driver cross the line if the race was originally scheduled to begin at 3.50 pm?

   ✓ Correct Ans A. 5:58 pm
   B. 4:33 pm
   C. 4:56 pm
   D. 4:34 pm

Explanation

The first driver used 1hr 12 min
The second driver used 1hr 12 min + 25 minutes = 1hr 37 mins
The third driver used 1hr 37 mins + 6 mins = 1hr 43 mins
If the race started 25minutes late from 3.50pm then it started at 3.50pm + 25mins = 4.15pm
The third-placed driver used 1hr 43mins, so he came in at
4.15pm + 1hr 43mins= 5.58pm

Q: 24 The Red Devils have won the League more than most teams in the League, even though they have not won the title in the last three years as The Saints have won consecutively in the last two years.

Tim: "The Red Devils are the best team in the League, having won the title 7 times in the last 10 years."
Patrick: "The Saints are the current Champions in the league."

If the information is true, whose reasoning is correct?

   A. Tim only

✓ Correct Ans **B. Patrick only**

   C. Both are wrong

   D. Both are right

Explanation

Patrick is right. On current form, The Saints are the best team in the League as they are the current Champions, even though The Red Devils were Champions for several years in the past.

Q: 25 Randall Kendham was a great middle-distance runner who first broke into national consciousness as an athlete at the University of Olympio. Then he broke the 1,500m National Records, which also was the National Collegiate Records. While the national record was broken nine years later, the Collegiate record of 3:40 mins stood for a long time until it was broken last year by an athlete from the University of Dolphin, who ran a semi-final race of the 2022 Collegiate Games at 3:29 mins.

What fraction of the original record time did the new record holders take off?

   A. $3/10$

   B. $6/7$

✓ Correct Ans **C. $1/20$**

   D. $1/30$

Explanation

The Kendham record was 3:40mins which is 220secs.
The new time is 3:29 minutes, which is 209 seconds. The difference is 11 seconds
The fraction of the old records taken off was: $11/220 = 1/20$.

Q: 26 There are three diners in a restaurant, having late meals. Two are regular and the other is just a casual customer at the restaurant. One had a meal of French fries, another hamburger and the third mashed potatoes.
The regulars decided to take the house coffee - one Spanish the other Irish as desserts, while the third man took ice cream.

After the meals, the man who ate fries left a large tip, the man who ate potatoes left a normal tip after his cup of Spanish coffee and the third man left no tip after his ice cream.

What did the man with the large tip drink?

    A. Spanish Coffee

✓ Correct Ans B. Irish Coffee

    C. Ice Cream

    D. Water

Explanation

The regulars drank coffee: one- Spanish; one –Irish and the casual drank Ice cream
The two coffee drinkers (regulars) left tips, and the ice cream drinker (casual) left no tip.
The Spanish coffee drinker (a regular) ate Potatoes, left a normal tip
The other coffee drinker left a large tip
And he took Irish coffee.

Q: 27 In a league competition, the 10 teams have to play each other at home and away. With three matches left, *The Park Rangers* sit atop the table with 51 points, the next-placed team has 43 points and the third-placed team has 41 points.

If a win attracts 3 points, a draw 1 point and a loss no point, what is the least possible position Park Ranger could finish the season with?

    A. First

✓ Correct Ans B. Second

    C. Third

    D. Fourth

Explanation

The next three matches fetch 9 points to a team who wins all 3.
The worst-case scenario for Park Rangers is to lose all three matches and its next two placed opponents winning theirs. So they end up as:
Park Rangers: 51 + 0 = 51 points
Second place: 43 + 9 = 52
Third-placed : 41 + 9 = 50
So the least possible position is second.

Q: 28 In a class of 30 students, Michael was in the top half of the class, Iris was in the top third of the class, Peggy's position was 16th and Ramos was among the first 5 brilliant students in the class.

How are these students placed in class, if all students are near end in their category?

    A. Michael, Iris, Peggy, Ramos
    B. Ramos, Michael, Iris, Peggy
✓ Correct Ans   C. Ramos, Iris, Michael. Peggy
    D. Michael, Ramos, Iris, Michael

Explanation

From the statement, Michael is in the first half which position within 1-15;
Iris is in the top third, which is between 1-10
Peggy is no 16 in the class,
Ramos is between 1-5
So their position:
Ramos 1 – 5
Iris: 1 -10
Michael 1 – 15
Peggy: 16

Q: 29 The chef uses only tender boneless meat to cook the special dish. There are 28 pieces of meat in the freezer. 12 are tough meat and the **rest** are tender.

If 7 of the **rest** are bony, how many pieces of the meat can be used for the special dish?

    A. 10
    B. 12
    C. 8
✓ Correct Ans   D. 9

Explanation

There are 28 pieces of meat, and 12 are tough so we have 28 – 12 = 16 tender pieces of meat.
Out of the 16, 7 are bony, so we can only use 16-7 = 9 pieces

Q: 30   *The year is made up of 4 seasons: Spring, Summer, Fall and Winter, each with different weather, temperature and climatic conditions.*

Dick: "Each year, the world undergoes four climatic changes."
Peter: "There are different weathers in a year."
Tom: "Some of the seasons' weathers are similar to each other."

If the information is true, whose **inference** is correct?

- A. Dick and Peter only
- B. Peter and Tom only
- C. Dick and Tom only
- ✓ Correct Ans D. None of the options provided

Explanation

Only Peter correctly infers that there are different weathers in a year. 'Peter only' is not among answer choices.

Q: 31   The following is the number of hats owned by 3 gentlemen.

- Leslie: black hats (4); white hats (1); brown hats (1)
- Patrick: black hats (2); white hats (2) brown hats (2)
- Arthur: black hats (4); white hats (1); brown hats (3)

If the value of Arthur's hats is worth $30 more than that of Leslie's, and a Black hat sells for $15 and a white hat sells for $20, what is the difference between the value of Patrick's hats and Arthur's?

- A. $45
- B. $35
- C. $40
- ✓ Correct Ans D. $25

Explanation

Between Arthur and Leslie : difference is 2 brown hats = $30;
1 brown hat = $15;
1 black hat = $15;
1 white hat = $20.

Patrick = 30 + 40 + 30 = 100
Arthur = 60 + 20 + 45 = 125

Difference = $25

Q: 32 Four friends, Smith, Beatrice, Fred, and Diana, each went to see a different movie: action drama, comedy, family drama, and thriller. You know that:
• Smith and Beatrice saw movies in somewhat similar genre.
• Fred did not see the comedy movie.
• Diana saw either the comedy or the thriller.

If Smith saw the family drama movie, what movie did Diana see?

- A. Action
- ✓ Correct Ans B. Comedy
- C. Drama
- D. Thriller

Explanation

Since Smith saw the family drama and Beatrice saw somewhat similar genre, Beatrice has seen action drama. This leaves comedy and thriller as possibilities for Diana. However, Fred didn't see the comedy movie, so Fred saw thriller, Diana must have seen the comedy.

Q: 33 A pirate map leads to a buried treasure on a deserted island. The map has four instructions:
• Turn right and walk for 5 paces.
• Turn right again and walk for 15 paces.
• Dig at the spot.

If you start facing east, where will you be digging for the treasure?

- A. North
- B. East
- ✓ Correct Ans C. South-West
- D. North-West

Explanation

Following the instructions:
You start facing east.
Turning right once puts you facing south.
Walking 5 paces south keeps you facing south.
Turning right again puts you facing west.
Walking 15 paces west keeps you facing west.
Digging at the spot means the treasure is buried directly in front of you, which is now South-west.

Q: 34 Three friends, Joy, Noah, and Olivia, each brought a different fruit to share at a picnic: apple, banana, and orange. You know that:
- Joy likes both apples and bananas.
- Noah does not like oranges.
- Olivia brought the fruit that Joy likes.

If Noah brought the apple, what fruit did Olivia bring?

A. Apple

✓ Correct Ans B. Banana

C. Orange

D. Impossible to determine

Explanation

While Joy likes both apples and bananas, Olivia brought the fruit Joy likes. If Noah brought the apple (leaving only orange and banana), and Joy likes apples, then it wouldn't make sense for Olivia to bring the orange. Therefore, Olivia must have brought the remaining fruit, which is the banana.

Q: 35 Three friends—Oliver, Penelope, and Quincy—embark on a quest through the Enchanted Forest. The forest has magical creatures and hidden paths with the following guiding principles:

- Principle A: "The path with fireflies is always taken before passing through the dense thicket."
- Principle B: "Once the crystal-clear pond is found, the journey must continue without encountering the mystical wolves."
- Principle C: "Talking owls are encountered immediately after passing through the dense thicket."

If Quincy encounters the talking owls, what can be inferred about the previous elements of the journey?

✓ Correct Ans A. The fireflies were encountered.

B. The crystal-clear pond was found.

C. The mystical wolves were bypassed.

D. Impossible to determine

Explanation

According to Principle C, talking owls are encountered immediately after passing through the dense thicket. Since Quincy encounters the talking owls, it implies that the previous element of the journey was the dense thicket, which means the fireflies were encountered before.

Q: 36 In the heart of the enchanted forest, there are three magical creatures: Gryphor, Mystique, and Zephyr. Each creature guards a different path leading to a hidden realm. The inscriptions near their lairs read:

• Gryphor: "Choose the path of courage, where the mightiest trees stand tall."
• Mystique: "Navigate through the shadows, where illusions conceal the way."
• Zephyr: "Follow the direction of the wind, where the air hums the secrets of the realm."

You have a compass and weather app on your smartphone. If you seek the hidden realm, which path should you choose?

    A. Gryphor's path
    B. Mystique's path
✓ Correct Ans C. Zephyr's path
    D. Impossible to determine

Explanation

Zephyr's inscription provides a clearer directive to follow the direction of the wind, suggesting a more straightforward path to the hidden realm. Direction of wind can be found with Compass and weather app on smartphone.

Q: 37 Which of the following is the correct figure?

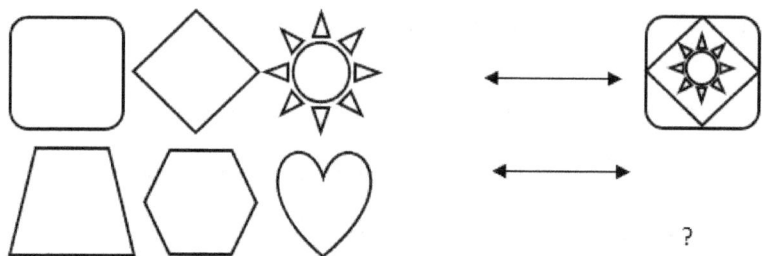

✓ Correct Ans A.

    B.

    C.

    D.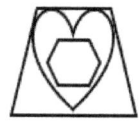

Explanation

It can be observed that in the first figure the rounded rectangle encloses the diamond and sun in the center. Thus, the same thing must be done with the second set of figure, in which the trapezoid will enclose the hexagon in its center. Option A is the correct answer.

Q: 38 In water, a word is shown as PRAKASH (mirrored). What is correct form of the word?

- A. PRASHAK
- ✓ Correct Ans B. PRAKASH
- C. PRSKHAP
- D. KRASHAP

Q: 39 Select the figure from the options which when placed in blank space would complete the pattern.

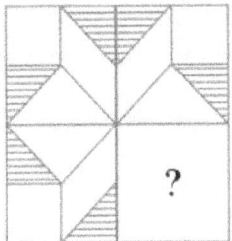

A.

✓ Correct Ans B.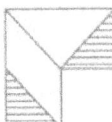

C.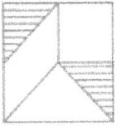

D.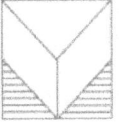

Explanation

Option (b) would complete the pattern.

Q: 40 Find the values of P and Q respectively in the given number pattern.

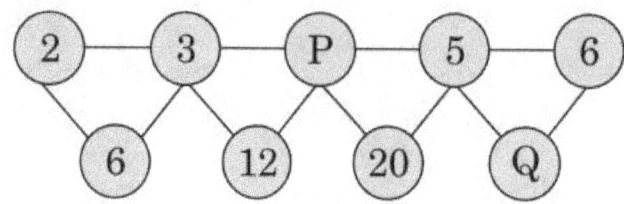

A. 5 and 28

✓ Correct Ans B. 4 and 30

C. 3 and 30

D. 2 and 18

Explanation

The pattern is as follows:

2   3   ④   5   6
 +1 +1 +1 +1

and   6   12   20   ㉚
       +6   +8   +10

Preptive Prepare Better
Series K Reading Comprehension Trial Test

Answers and Solutions

Q: 1 What does Extract B reveal about the passengers and crew of the Titanic?

- A. Their collective disregard for safety protocols
- ✓ Correct Ans **B. Their diversity in social and economic backgrounds**
- C. Their advanced knowledge of maritime navigation
- D. Their unanimous decision to abandon ship

Explanation

Extract B sheds light on the diverse backgrounds of the passengers and crew aboard the Titanic, emphasizing the varied experiences and stories of individuals from different social and economic backgrounds.

- A. They serve as evidence of conspiracy theories.
- ✓ Correct Ans **B. They fuel ongoing debates among historians.**
- C. They detract from the significance of the event.
- D. They are dismissed as irrelevant to understanding the disaster.

Explanation

Extract B suggests that unanswered questions surrounding the Titanic disaster fuel ongoing debates among historians and researchers, contributing to the complexity of understanding the event.

Q: 3 How does Extract A characterize the sinking of the Titanic in terms of human nature?

- A. As a display of unwavering courage and resilience
- ✓ Correct Ans **B. As a reflection of human arrogance and humility**
- C. As a testament to human greed and corruption
- D. As a result of external factors beyond human control

Explanation

Extract A portrays the sinking of the Titanic as a reflection of human arrogance, demonstrated through complacency and hubris, as well as the importance of humility and vigilance in the face of nature's unpredictability.

Q: 4 What aspect of the Titanic's legacy does Extract B primarily seek to explore?

- A. Its lasting impact on maritime technology
- B. Its influence on popular culture and media
- ✓ Correct Ans C. Its untold stories and lesser-known facets
- D. Its role in shaping historical narratives

Explanation

Extract B primarily aims to explore the untold stories and overlooked aspects of the Titanic's legacy, offering new perspectives and deeper insights into the monumental event.

Q: 5 According to Extract A, what major reforms were implemented in response to the Titanic disaster?

- A. Stricter regulations on passenger accommodations
- B. Enhanced training for maritime crews
- ✓ Correct Ans C. Improved communication systems and navigation procedures
- D. Increased reliance on luxury amenities

Explanation

Extract A discusses that significant reforms were implemented post-Titanic disaster, including improvements in communication systems and navigation procedures, aimed at enhancing maritime safety and preventing similar tragedies in the future.

Q: 6 According to Extract B, what is a significant challenge faced by African countries due to rapid urbanization?

- A. Decreased demand for food
- B. Increased availability of arable land
- ✓ Correct Ans C. Higher dependence on food imports
- D. Improved access to agricultural resources

Explanation

Extract B discusses how rapid urbanization in African countries increases demand for food while reducing the availability of arable land, leading to higher dependence on food imports and vulnerability to global market fluctuations.

Q: 7 Which concept is central to the discussion of food security in both extracts?

- A. Economic inequality
- B. Malnutrition
- ✓ Correct Ans C. Resilience
- D. Urbanization

Explanation

Both extracts highlight resilience as a central concept in ensuring food security, emphasizing the importance of building food systems that can withstand shocks and stresses.

Q: 8 According to both extracts, what is critical for fostering resilient and sustainable communities?

- A. Increased food imports
- B. Promotion of monoculture farming
- ✓ Correct Ans C. Investment in rural development
- D. Dependence on global markets

Explanation

Both extracts emphasize that investment in rural development is critical for fostering resilient and sustainable communities by enhancing agricultural productivity and food security.

Q: 9 How does Extract B characterize the impact of conflict and political instability on food security?

- A. They enhance food production and distribution.
- B. They have minimal effect on food availability.
- ✓ Correct Ans C. They disrupt food production and access.
- D. They facilitate efficient food aid delivery.

Explanation

Extract B discusses how conflict and political instability disrupt food production, distribution, and access, leading to food shortages and humanitarian crises.

Q: 10 Which extract highlights roofing known for its sophisticated appearance crafted from natural stone?

- A. Extract A
- ✓ Correct Ans B. Extract B
- C. Extract C
- D. Extract D

Explanation

The correct answer is (b) Extract B, which talks about slate roofing that boasts elegance and durability, being quarried from natural stone and offering sophisticated appeal.

Q: 11 Which extract mentions roofing that is particularly renowned for its resistance to harsh weather conditions and exceptional longevity?

✓ Correct Ans **A.** Extract A
B. Extract B
C. Extract C
D. Extract D

Explanation

The correct answer is (a) Extract A, as it discusses metal roofing known for durability, sleek appearance, and exceptional resistance to harsh weather conditions.

Q: 12 Which extract discusses roofing that offers a balance between cost-effectiveness, diverse styles, and reliable protection against moderate weather conditions?

A. Extract A
B. Extract B
✓ Correct Ans **C.** Extract C
D. Extract D

Explanation

The correct answer is (c) Extract C, which describes asphalt shingles as a popular and cost-effective choice offering diverse styles, ease of installation, affordability, and reliable protection against moderate weather conditions.

Q: 13 Which extract mentions roofing options that cater to diverse aesthetic preferences through various styles like standing seam and metal tiles?

✓ Correct Ans **A.** Extract A
B. Extract B
C. Extract C
D. Extract D

Explanation

The correct answer is (a) Extract A, which talks about metal roofing and its various styles like standing seam and metal tiles, catering to diverse aesthetic preferences.

Q: 14 Which extract highlights roofing known for its labor-intensive installation but cherished for its enduring sophistication?

- A. Extract A
- ✓ Correct Ans **B. Extract B**
- C. Extract C
- D. Extract D

Explanation

The correct answer is (b) Extract B, which discusses slate roofing that is labor-intensive to install but cherished for its enduring sophistication and natural beauty.

Q: 15 What is the primary theme of Emily Dickinson's poem "The Chariot"?

- A. Celebration of life
- B. Fear of death
- ✓ Correct Ans **C. Acceptance of mortality**
- D. Joy of youth

Explanation

The poem primarily revolves around the acceptance of mortality and the journey toward death.

Q: 16 What is the significance of the carriage holding "just ourselves And Immortality"?

- A. It suggests the speaker's fear of death
- ✓ Correct Ans **B. It represents the eternity of the journey**
- C. It implies that the journey is lonely
- D. It symbolizes the inevitability of death

Explanation

The carriage holding "just ourselves And Immortality" implies that the journey with Death extends into eternity.

Q: 17 What is the deeper meaning behind the line, "The roof was scarcely visible, The cornice but a mound"?

- A. The speaker is describing a mansion
- ✓ Correct Ans **B. The house represents a grave**
- C. It depicts a house sinking into the ground
- D. It signifies the passing of time

Explanation

The description of the house with a barely visible roof and a cornice as a mound symbolizes a grave.

Q: 18 What emotion does the speaker convey through the line, "Since then 't is centuries; but each / Feels shorter than the day"?

- A. Regret
- B. Fear
- C. Longing
- ✓ Correct Ans D. Perceived passage of time

Explanation

The writer perceives that centuries have passed, but they feel shorter than a single day, indicating a reflection on the passage of time.

Q: 19 What is the primary emotion expressed by the speaker in the poem?

- A. Joy and celebration
- ✓ Correct Ans B. Sorrow and longing
- C. Excitement and anticipation
- D. Peace and contentment

Explanation

The poem "Break, Break, Break" primarily expresses the speaker's deep sense of sorrow and longing, evident in the repeated lamentation for something lost.

Q: 20 What is the poet's lamentation about in the poem?

- A. The joy of playing by the sea
- B. The absence of fishermen
- C. The loss of his sister's voice
- ✓ Correct Ans D. The inability to express his thoughts

Explanation

The speaker in the poem expresses a desire for his tongue to articulate the thoughts and emotions that arise within him, indicating a yearning to communicate something deeply felt but unspoken.

Q: 21 What does the phrase "the touch of a vanished hand" suggest in the poem?

        A. The sensation of cold water

✓ Correct Ans B. The longing for lost affection

        C. The joy of sailing on the bay

        D. The imagery of a vanished ship

Explanation

The phrase "the touch of a vanished hand" conveys a strong sense of longing and grief for the loss of affection or connection, emphasizing the absence of a loved one.

Q: 22 What is the deeper meaning suggested by the contrast between the activities of the fisherman's boy and the sailor lad?

        A. The speaker's admiration for their adventurous spirit

        B. The speaker's envy of their carefree lives

✓ Correct Ans C. The juxtaposition of joyous activities against the speaker's sorrow

        D. The speaker's longing for the sea and its delights

Explanation

The contrast between the carefree activities of the fisherman's boy and the sailor lad emphasizes the speaker's sense of sorrowful isolation, contrasting their joyous lives with the speaker's grief.

Q: 23 What is the symbolic significance of the sea in the poem?

        A. It represents the speaker's love for sailing

        B. It symbolizes the vastness of nature

✓ Correct Ans C. It signifies the passage of time and life's impermanence

        D. It embodies the speaker's hope for the future

Explanation

The sea symbolizes the relentless passage of time and the impermanence of life, as reflected in the ceaseless breaking of waves against the shore.

Q: 24 Which one fits in (1) ?

        A. Sentence A

        B. Sentence B

✓ Correct Ans C. Sentence G

        D. Sentence H

Explanation

The first paragraph presents the introduction about importance of school system. Therefore, "C" is the answer.

Q: 25 Which one fits in (2) ?

✓ Correct Ans **A.** Sentence A
B. Sentence B
C. Sentence C
D. Sentence D

Explanation

The prior sentences are about the subjects the school system has. Therefore, the correct answer should be "a"

Q: 26 Which one fits in (3) ?

A. Sentence A
B. Sentence D
✓ Correct Ans **C.** Sentence B
D. Sentence E

Explanation

The next line has the word "skills" and the answer "c" has some skills that the school possesses.

Q: 27 Which one fits in (4) ?

A. Sentence A
B. Sentence B
C. Sentence H
✓ Correct Ans **D.** Sentence E

Explanation

The paragraph mentions the opportunities for interactions and building up good relationships. Therefore, the sentence should be "d" which describes the promotion of socialization and a sense of community.

Q: 28 Which one fits in (5) ?

A. Sentence A
B. Sentence B
✓ Correct Ans **C.** Sentence C
D. Sentence D

Explanation

The next lines of the paragraph mention the values hathe students should have. The correct answer "c" also shows the way how school helps in developing them.

Q: 29 Which one fits in (6) ?

- A. Sentence A
- B. Sentence B
- C. Sentence C
- ✓ Correct Ans D. Sentence D

Explanation

The paragraph mentions about the inequalities. Thus, correct answer should be "d"

Q: 30 Which one fits in (7) ?

- A. Sentence A
- B. Sentence G
- ✓ Correct Ans C. Sentence H
- D. Sentence C

Explanation

The paragraph mentions about the technology. So, the correct answer is "c"

Preptive Prepare Better
Series K Mathematical Reasoning Trial Test

Answers and Solutions

Q: 1 Joy and Siti had a total of 360 beads at first. Joy lost 28 beads while Siti bought another 18 beads. Both of them had an equal number of beads in the end.

How many beads did **Joy have in the end?**

    A. 157

✓ Correct Ans B. 175

    C. 185

    D. 203

Explanation

<u>At Start :</u> Joy had n, Siti had 360 - n.
<u>At End</u> : Joy had n-28, Siti had 378 - n.

n - 28 = 378 - n; nx2 = 378+28 = 406
n = 203
Joy had n-28 at end i.e. 175 at end.

Q: 2 The location of three cities A B C is shown in the picture. The distance between A and B is 45 km. The distance between A and C is 161 km.
What is the distance between B and C?

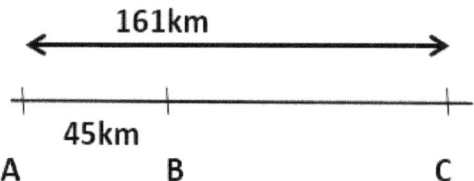

✓ Correct Ans A. 116km

    B. 45 km

    C. 134km

    D. 125km

Explanation

Let's subtract the distance between A and B from the distance between A and C. Then we get the distance between B and C.
161km – 45km = 116km
The correct answer is: A. 116km

Q: 3 Jin and her brother had some marbles. The ratio of the number of marbles Jin had to the number of marbles that her brother had was 1 : 4 at first. After their mother gave Jin 54 more marbles, Jin and her brother had the same number of marbles.
How many marbles did Jin's brother have at first?

✓ Correct Ans **A.** 72 marbles

B. 18 marbles

C. 54 marbles

D. 70 marbles

Explanation

At start : Jin = n, Brother = 4n
At end : Jin = n+54 , Brother = 4n

4n = n + 54 ; 3n = 54 ; n = 18
Brother had 4n = 72 marbles at first.

Q: 4 Lisa had four times as much money as Joshua. After their mother gave each of them an equal amount of money, Lisa had thrice as much money as Joshua.
If Joshua had $42 in the end, how much money did Lisa have at first?

✓ Correct Ans **A.** $112

B. $84

C. $28

D. $42

Explanation

Start : Lisa = 4n, Joshua = n
Mother gave w amount of money.
End : Lisa = 4n + w, Joshua = n+w

n + w = 42 ...(1)
4n + w = (n+w)x3 = 126 ..(2)

(2) - (1) :
3n = 126 - 42 = 84; n = 28
4n = 28x4 = 112

Q: 5 There were $3/7$ as many tables as chairs in a school hall. Mr. Chan added 60 tables and 60 chairs into the school hall for an event.
As a result, the number of tables was $3/5$ the number of chairs.
How many chairs were in the hall in the end?

A. 100

✓ Correct Ans B. 200

C. 220

D. 250

Explanation

Start : Tables = 3n, Chairs = 7n
End : Tables = 3n+60, Chairs = 7n+60

$3n + 60 = 3/5 (7n+60)$
$15n + 300 = 21n + 180$
$6n = 120; n = 20$
Chairs at end = 7n+60 = 140+60 = 200

Q: 6 The bar chart shows the number of students who took a variety of languages.

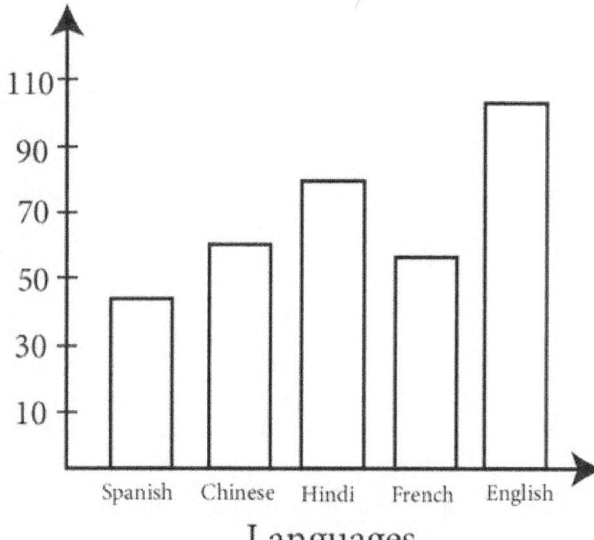

Which of the following is a correct statement?

A. Students who took French and Chinese have the same amount.

✓ Correct Ans B. The difference between the students who took English and Hindi is 20

C. The total of students who took English, Hindi, and, French is 640.

D. The students who took Spanish are more than those who took French.

Explanation

A. Students who took French and Chinese have the same number. Wrong. Chinese has more students than French.
B. The difference between the students who took English and Hindi is 20. Correct. 100 − 80 = 20
C. The total of students who took English, Hindi, and, French is 640. Wrong. 100 + 80 + 60 = 240.
D. The students who took Spanish are more than those who took French. Wrong. The students who took Spanish are less than the students who took French.

Q: 7 The line chart below shows the scores A in English in a variety of classes.

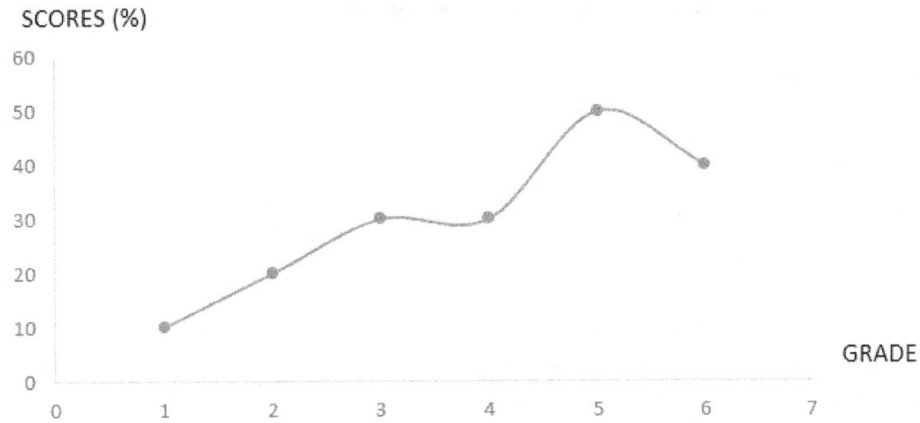

Which of the following is a correct statement?

    A. The total scores of Grades 6 and 3 are 30.

✓ Correct Ans   **B.** The total scores of Grades 1, 4, and 5 are 90

    C. The difference scores between Grades 2 and 1 is 45

    D. The difference scores between Grades 5 and 3 is 60.

Explanation

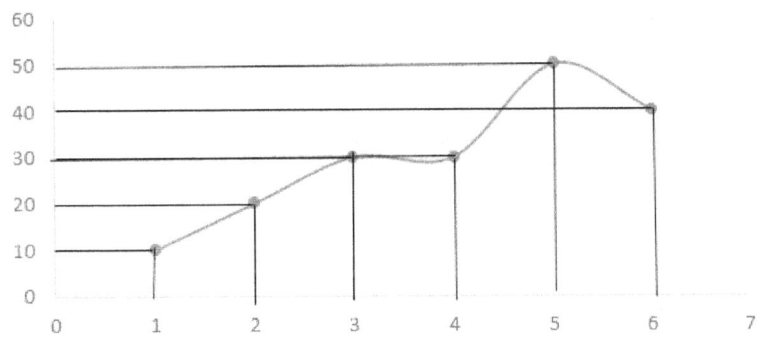

A. The total scores of Grades 6 and 3 are 30. Wrong. 40 + 30 = 70
B. Grades 1, 4, and 5 total scores are 90. Correct. 10+30+50 = 90
C. The difference scores between Grades 2 and 1 is 45. Wrong. 20-10=10
D. The difference scores between Grades 5 and 3 is 60. Wrong. 50-30=20

Q: 8 Pinky Co makes pink paint by mixing red paint and white paint in a ratio of 3:4. Slacky Co makes pink paint by mixing red paint and white paint in a ratio of 5:7.
Which company uses a higher proportion of red paint in their mixture?

- A. They are the same
- ✓ Correct Ans B. Pinky Co
- C. Slacky Co
- D. It is impossible to tell

Explanation

The proportion of red paint for Pinky Co is $3/7$. The proportion of red paint for Slacky Co is $5/12$. We can compare fractions by putting them over a common denominator using equivalent fractions. $3/7 = 36/84$ and $5/12 = 35/84$. So, $3/7$ is a bigger fraction. Pinky Co uses a higher proportion of red paint.

Q: 9 Below shows a bar chart.

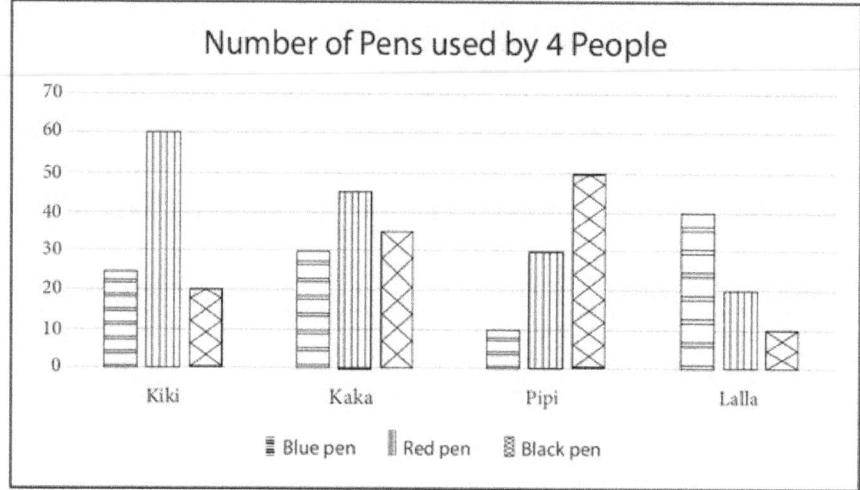

Which of the following is a correct statement?

- A. Kiki use more blue pen than Lalla
- B. Kaka uses more red pen than Kiki.
- ✓ Correct Ans C. Pipi uses more pens than Lalla.
- D. Kaka uses more black pens than Pipi.

Explanation

A. Kiki uses more blue pens than Lalla. Wrong. Kiki uses less blue pen than Lalla.
B. Kaka uses more red pen than Kiki. Wrong. Kaka uses less red pen than Kiki.
C. Pipi uses more pens than Lalla. Correct. 10+30+50 = 90 pens used by Pipi. 40+20+10=70 pens used by Lalla.
D. Kaka uses more black pens than Pipi. Wrong. Kaka uses less black pen than Pipi.

Q: 10 The picture shows a graph.

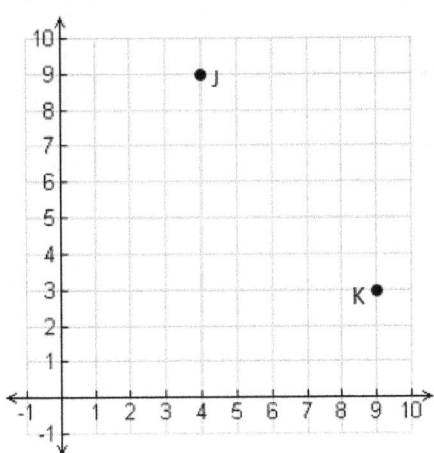

Points J, K, and M form a right-angled triangle. State the possible coordinates of M from the answer options.

✓ Correct Ans **A.** (4,3)
B. (5,3)
C. (6,9)
D. (9,3)

Explanation

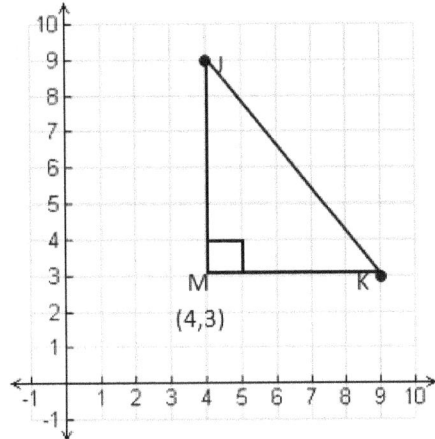

A Cartesian plane can be defined as a plane formed by the intersection of two coordinate axes that are perpendicular to each other. The horizontal axis is called the x-axis and the vertical one is the y-axis. These axes intersect with each other at the origin whose location is given as (0, 0). Any point on the cartesian plane is represented in the form of (x, y). Here, x is the distance of the point from the y-axis and y is the distance from the x-axis.

Q: 11 A box $\frac{1}{2}$ filled with beans has a total mass of $8\frac{2}{5}$ kg.

When the same box is completely filled with beans, the total mass is $15\frac{8}{10}$ kg.
What is the mass of the box when it is empty?

✓ Correct Ans **A.** 1 kg

B. 3.56 kg

C. 3.9 kg

D. 1.2 kg

Explanation

Box + Bean/2 = $\frac{42}{5}$ .. (1)
Box + Bean = $\frac{158}{10}$ .. (2)

(1) x 2 - (2) :
Box x 2 + Bean = $\frac{84}{5}$ = $\frac{168}{10}$
Box + Bean = $\frac{158}{10}$

Box = $\frac{10}{10}$ = 1 kg

Q: 12 A total of 32,280 people took part in the marathon. There were 5 times as many adults as children. When 30 women and 30 children withdrew from the marathon, the number of women who took part was twice the number of children who took part. Find the number of men who took part in the marathon.

A. 55,256

B. 1,190

C. 16,550

✓ Correct Ans **D.** 16,170

Explanation

At first : 6 parts, 1 part children
Children = 32280/6 = 5380
Adults = 5380 x 5 = 26900

At end :
Adults = 26900 - 30 = 26870
Children = 5380 - 30 = 5350
Women = 10700
Men = 26870 - 10700 = 16170

Q: 13 The diagram below shows a line graph.

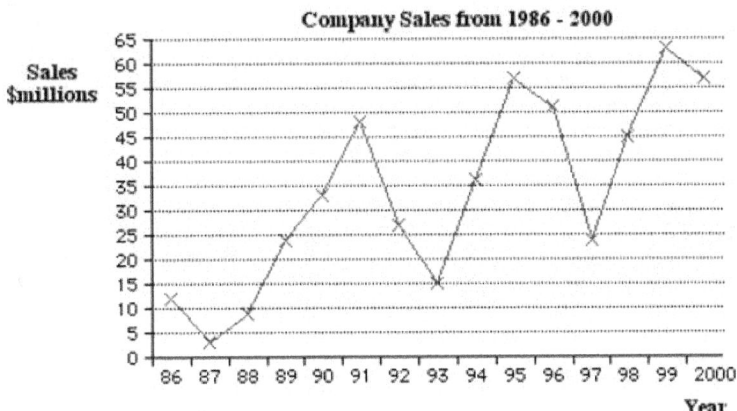

Which of the following is a correct statement?

- A. The sales in 1990 were lesser than the sales in 1993.
- B. The sales from 1991 to 1993 show an increasing pattern.
- ✓ Correct Ans C. The sales from 1997 to 1999 show an increasing pattern.
- D. The sales in 1995 were higher than the sales in 1999.

Explanation

A. The sales in 1990 were lower than the sales in 1993. Wrong. The sales in 1990 were more than the sales in 1993.
B. The sales from 1991 to 1993 show an increasing pattern. Wrong. The sales from 1991 to 1993 show a decreasing pattern.
C. The sales from 1997 to 1999 show an increasing pattern. Correct
D. The sales in 1995 were higher than the sales in 1999. Wrong. The sales in 1995 were lower than the sales in 1999.

Q: 14   The diagram below shows a net of a cuboid.

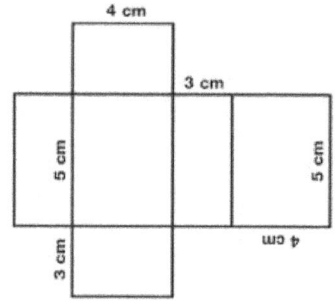

What is the volume of the cuboid?

A. 50cm³

✓ Correct Ans B. 60cm³

C. 30cm³

D. 55cm³

Explanation

Volume: width x length x height = 5 x 4 x 3 = 60cm³

?

Q: 15   The diagram below shows two rectangles.

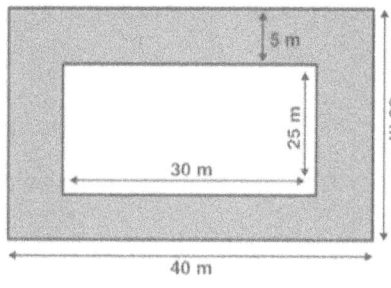 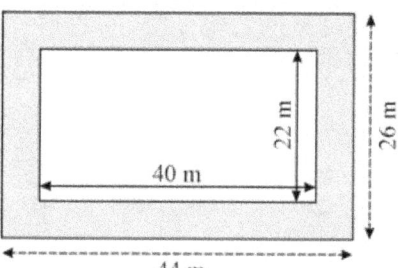

Find the **total** shaded area.

A. 245m²

B. 945m²

C. 945m²

✓ Correct Ans D. 914m²

Explanation

Area of Rectangle = length × width
(40m x 35m) – (30m x 25m) = 1,400m - 750m = 650m²
(44m x 26m) – (40m x 22m) = 1,144m – 880 = 264m²
Total: 914m²

Q: 16 James prepared slices of mini cakes to be served during a party. When he gives 7 slices to each guest, he will have 2 slices left. When he gives 3 slices to each guest, he will have 134 slices left.
How many guests were at the party?

☑ Correct Ans **A.** 33
B. 32
C. 43
D. 35

Explanation
There were 'n' guests.
7xn + 2 = 3xn + 134
4xn = 132; n = 33

?

Q: 17 JK : Today, my age is 55 years old.
RM: 36 days later, my age will be the same as yours.
RJ : Oh, today is 6 March 2019!

According to the conversation above, what is RM's date of birth ?

☑ Correct Ans **A.** 10th April 1964
B. 1st April 1954
C. 14th April 1955
D. 22nd April 1964

Explanation
31 days in March − 5 days = 26 days
36 − 26 = 10 days = 10th April
2019 − 55 = 1964
So, 10th April 1964

Q: 18

Shamila had some cookies. She sold $\frac{3}{4}$ of the cookies on the first day. She sold $\frac{1}{2}$ of the remaining cookies plus 2 more on the second day. She had 84 cookies left in the end.
How many cookies did Shamila have at first?

- A. 172
- B. 688 ✓ Correct Ans
- C. 86
- D. 700

Explanation

Use working backwards strategy.
$1/2$ of remaining = 2 + 84 = 86
Remaining = 86 x 2 = 172
$1/4$ of total = 172
$4/4$ of total = 172 x 4 = 688

Q: 19

During a trivia quiz, points were awarded for questions answered as shown below.

Correct	5 Points
Wrong	-2 Points
Missed(Unanswered)	-1 Points

For Janice, the ratio of the questions answered correctly to questions answered wrongly to questions that were missed out was 9 : 2 : 1.

If Janice was awarded 360 points, how many questions are there in all?

- A. 80
- B. 100
- C. 110
- D. 108 ✓ Correct Ans

Explanation

C : W : M = 9 : 2 : 1
Total: 12 questions in a group.

Let us calculate point of a group.
9 x 5 = 45; 2 X 2 = 4
1 x 1 = 1; 4 + 1 = 5
45 - 5 = 40

How many groups are there ?
360 ÷ 40 = 9
Total questions = 9 x 12 = 108

Q: 20 Find the missing number based on the figure below.

?			
11	9		
4	7	9	
2	3	7	8

A. 10

B. 11

C. 30

✓ Correct Ans D. 20

Explanation

The sum of every row is 20.
2nd row = 11+9 = 20
3rd row = 4+7+9 = 20
4th row = 2+3+7+8 = 20
So we can assume 1st row is 20.

Q: 21 Amir started cycling from his house to town A at 9:00 a.m.

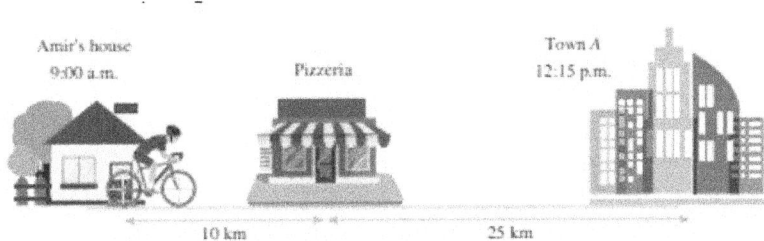

Along the way, he passed a pizzeria but didn't stop. He arrived in town A at 12:15 p.m. Calculate the average speed of the journey in km/h.

A. 13.78 km/h

✓ Correct Ans B. 10.77 km/h

C. 18.77 km/h

D. 17.87 km/h

Explanation

Total distance = 10 km + 25 km = 35 km
Total time taken = 12:15 – 9:00 = 3 h 15 min
Average speed = 35 km / 3.25 h = 10.77 km/h

Q: 22 Jamie has some 50¢ coins and Kumar has some $1 coins. Jamie has 12 more coins than Kumar. The total amount of money Kumar has is $15 more than the total amount of money Jamie has. How many coins does Kumar have?

    A. 46
    B. 36
    C. 24
✓ Correct Ans D. 42

Explanation

Jamie has 'n+12' coins, Kumar has 'n' coins.
Jamie's value = 0.5 x (n+12)
Kumar's value = 1 x n

Value difference = n - 0.5n - 6 = 0.5n - 6 = 15
0.5 n = 21; n = 42

Q: 23 Some identical lightbulbs were hung on a rod 91 cm long. The first and the last lightbulb were hung 14 cm away from each end of the rod. The rest of the lightbulbs were hung at an equal distance of 9 cm apart.
How many lightbulbs were hung on the rod altogether?

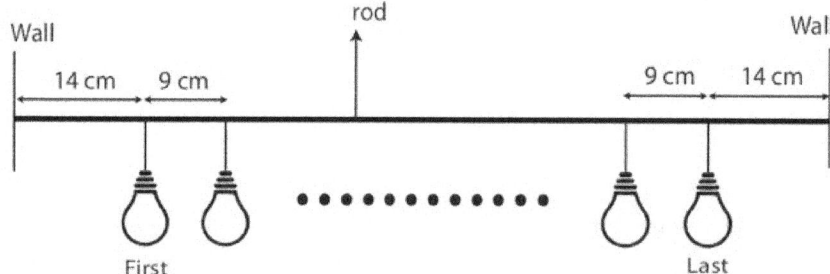

    A. 4
    B. 5
    C. 7
✓ Correct Ans D. 8

Explanation

91 - 14 - 14 = 63
63 ÷ 9 = 7
7 + 1 = 8
Ans: 8 bulbs

Q: 24 The graph shows the attendance of students to the school and average marks they got.

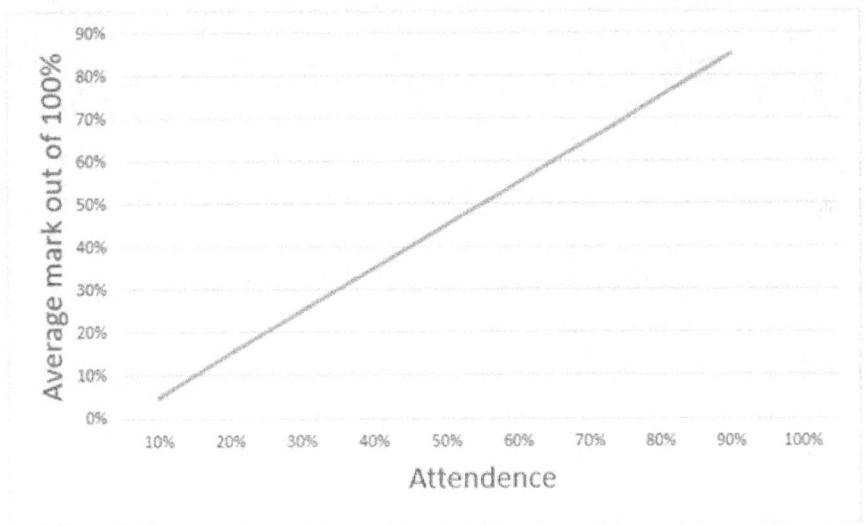

What is the average mark of the students who have an attendance of 40%?

A. 65%

B. 55%

C. 15%

✓ Correct Ans D. 35%

Explanation

Find 40% from x axis and move upwards until you intersect the graph. Now move to the left until you intersect y axis. You intersect y axis at 35.

Q: 25 Which of these statements is/are correct?

X. $2 - \frac{5}{9}$ is less than $1\frac{1}{3}$
Y. $\frac{1}{4} + \frac{3}{4}$ is equal to 1
Z. $\frac{5}{6}$ is less than $\frac{4}{3}$

- A. X and Y only
- ✓ Correct Ans B. Y and Z only
- C. Y only
- D. Z only

Explanation

X. $2 - \frac{5}{9}$
$= \frac{18}{9} - \frac{5}{9} = \frac{13}{9}$

$1\frac{1}{3} = \frac{4}{3} = \frac{12}{9}$
X is incorrect.

Y. $\frac{1}{4} + \frac{3}{4} = \frac{4}{4} = 1$
Y is correct.

Z. $\frac{4}{3} = \frac{8}{6}$
Z is correct.

Q: 26 A resident of Los Angeles took a vacation to New Jersey. She leaves Los Angeles at 1400 on Tuesday for a 3-hour flight to New York. She spends 12 hours in New York before catching a 2-hour flight to New Jersey. If she spent two days in New Jersey, when did she leave for Los Angeles?

- A. 0500 Thursday
- B. 0500 Friday
- C. 0700 Thursday
- ✓ Correct Ans D. 0700 Friday

Explanation

14:00 +3 hours = 1700 hrs, when she arrived in New York.
1700 + 12 hrs. = 29:00 – 24 hrs. = 0500 hrs. Wednesday is when the resident leaves New York.
She takes 2 hours to arrive in New Jersey and reaches the city at 0500 + 2 hrs. = 0700 hrs.
Due to the 2-day stay, she left for Los Angeles at 0700 hrs. Friday.

Q: 27 Every symbol represents a particular number and works as follows : each star represents 3; hence 3 x 3 =9.

| ★ | ★ | 9 |

What's the largest number among the three numbers (**including one that needs to be calculated**) in the following case?

☀	☀	36
☁	☁	64
☀	☁	

A. 100

✓ Correct Ans B. 64

C. 36

D. 128

Explanation

So, each sun represents 6 to get 36 from 6 x 6. On the other hand, each cloud represents 8 hence 64 from 8 x 8. So, the sun x cloud is 8 x 6 = 48.
Therefore, the largest is 64.

?

Q: 28 Below is a set of 3 jugs. If jug Z has a capacity of 800 ml, how far will it be from being full (**remaining empty capacity**) after ½ of what's in jug X and $1/5$ of the content in jug Y is poured into it?

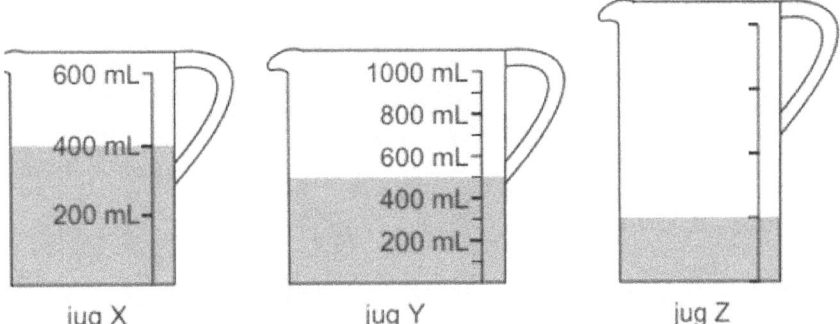

A. 200 ml

B. 100 ml

C. 500 ml

✓ Correct Ans D. 300 ml

Explanation

Currently, jug Z is ¼ full, and that's ¼ x 800 = 200 ml. Jug X reads 400 ml, and ½ of that is 200 ml. Lastly, jug Y reads 500 ml, and 1/5 is 100 ml. So, the final reading in jug Z will be 200 + 200 + 100 = 500 ml. What's remaining for it to be full is 800 − 500 = 300 ml.

Q: 29 Lenny has been reading a book that's 600 pages long for several days now. On the first day, she just read 50 pages. The next day, she read a quarter of the whole book. She read ½ of the remaining pages today.
So, how many more pages to go?

A. 550
B. 150
C. 400
✓ Correct Ans D. 200

Explanation

If she read 50 pages, the remaining pages were 600 – 50 = 550. She later read a quarter of the book, which is ¼ x 600 = 150 pages. Half of the remaining pages = ½ (600 – (50+150)) = 200. 50 +150 + 200 = 400. So, there are 600 – 400 = 200 pages left.

Q: 30 Ken and Ron had some stickers. Ken gave Ron $1/4$ of his stickers. As a result, Ken had 51 stickers and Ron had 62 stickers in the end.
How many stickers did Ron have at first?

A. 38
B. 42
C. 40
✓ Correct Ans D. 45

Explanation

Fraction of Ken's stickers left → 1 – $1/4$ = $3/4$
$3/4$ of Ken's stickers → 51
$1/4$ of Ken's stickers → 51 ÷ 3 = 17
Ken gave Ron 17 stickers.
Ron's stickers before receiving the 17 stickers from Ken → 62 – 17 = 45
Ron had 45 stickers at first.

Q: 31 Chloe and Sam had an equal number of apples at first. Chloe bought 3 more apples and Sam bought 23 more apples. Chloe then had 60% as many apples as Sam.
How many apples did Sam have at first?

✓ Correct Ans A. 27
B. 23
C. 25
D. 28

Explanation

Start : Chole = n, Sam = n
End : Chole = n+3, Sam = n+23

60% = $3/5$
n+3 = $3/5$ (n+23)
5n + 15 = 3n + 69; 2n = 69 - 15 = 54; n = 27

Q: 32 Teachers at Marybrown High School asked students to list their favorite desserts.

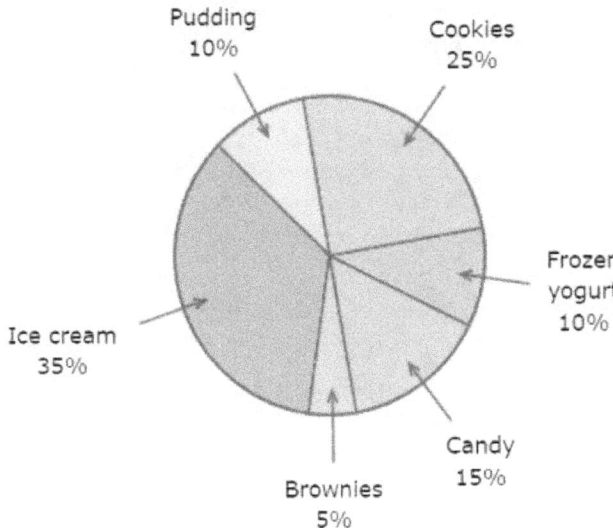

What is the measure of the central angle in the "Cookies" section?

A. 45°

✓ Correct Ans B. 90°

C. 120°

D. 150°

Explanation

Based on the graph, 25% of votes were for cookies.
Find 25% of 360°
25% of 360° = 0.25 x 360° = 90°

Q: 33 A cold water and hot water tap supply water to the bathtub at the rate of 45 liters per minute and 15 liters per minute respectively.
The drain empties water at the rate of 12 liters per minute.
The volume of the tub is 624 liters.

When all the taps and drains are turned on at the same time (**tub was empty before**), how long will it take to overflow in minutes?

A. 10 mins

B. 12 mins

✓ Correct Ans C. 13 mins

D. 15 mins

Explanation

The increase in volume = 45 + 15 − 12 = 48
Time after which it will overflow = $624/48$ = 13 minutes

Q: 34 Karen had some muffins. She gave $1/6$ of them to Mary, 32 to Amy, and kept the remaining 13 muffins for herself. How many muffins did Karen have at first?

- A. 48
- B. 45
- C. 52
- ✓ Correct Ans D. 54

Explanation

32 + 13 = 45
6/6 − 1/6 = 5/6
5/6 of the muffins → 45
1/6 of the muffins → 45 ÷ 5 = 9
Muffins at first → 6 × 9 = 54
Karen had 54 muffins at first.

Q: 35 10 pupils were standing in a straight row at equal distances apart from each other. If the 2nd pupil was 20 m away from the 6th pupil, what was the distance between the first pupil and the last pupil?

- A. 40m
- B. 56m
- ✓ Correct Ans C. 45m
- D. 42m

Explanation

Number of gaps between 2nd and 6th pupil → 6 − 2 = 4
20 m ÷ 4 = 5 m (Length of each gap)
10 − 1 = 9 (Total number of gaps)
9 × 5 m = 45 m
The distance between the first pupil and the last pupil was 45 m

Preptive Prepare Better
Series K Thinking Skill Trial Test

Answers and Solutions

Q: 1 Detective Jackson investigates a theft at a museum. Three suspects, Amelia, Ben, and Chloe, provide clues:

Amelia: "I was at the opera with Ben."
Ben: "I was at the opera with Amelia."
Chloe: "I was at the library all evening."

Later, CCTV footage reveals only one culprit. It is not known if the culprit is one of the suspects. If both Amelia and Ben are telling the truth, who must be the thief?

- A. Amelia
- B. Ben
- C. Chloe
- ✓ Correct Ans D. Impossible to determine

Explanation

If both Amelia and Ben are truthful, they were both at the opera together. This eliminates both of them as suspects. However, it doesn't guarantee Chloe's innocence, as another person could have committed the theft while she was at the library. Therefore, without further information, it's impossible to determine the thief based solely on the clues and camera footage.

Q: 2 A scientist studying a newly discovered element, "Element X," makes the following observations:

- No object is both magnetic and radioactive.
- All Element X samples are radioactive.

Based on these observations, which of the following is the most likely characteristic of Element X?

- ✓ Correct Ans A. Element X cannot be magnetic.
- B. Element X will always remain radioactive.
- C. Only Element X is radioactive.
- D. The radioactive property of Element X can be neutralized.

Explanation

While the second statement confirms all Element X samples are radioactive, it doesn't imply they possess other properties like magnetism. Conversely, the first statement tells us that no object can be both magnetic and radioactive. Combining these, the most likely conclusion is that Element X itself cannot be magnetic.

Q: 3 A bakery receives orders for three types of muffins: blueberry, cranberry, and chocolate chip. They process orders through a machine that follows these rules:

- Any order with "blueberry" in it outputs "breakfast muffin."
- Any order without "blueberry" but with "cranberry" or "chocolate chip" outputs "snack muffin."
- Any order without "blueberry," "cranberry," or "chocolate chip" outputs "plain muffin."

What would be the output for the following orders?

- Order 1: Blueberry and Cranberry
- Order 2: Chocolate Chip only
- Order 3: None of the known ingredients
- Order 4: Apple Fritter only

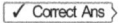 **A.** Breakfast muffin, Snack muffin, Plain muffin, Plain muffin

B. Snack muffin, Breakfast muffin, Plain muffin, Plain muffin

C. Breakfast muffin, Breakfast muffin, Plain muffin, Plain muffin

D. Snack muffin, Snack muffin, Plain muffin, Plain muffin

Explanation

Order 1: Contains both "blueberry" and "cranberry," so rule 1 applies. Output: Breakfast muffin.
Order 2: Contains "chocolate chip," so rule 2 applies. Output: Snack muffin.
Order 3: Doesn't contain any specified ingredients, so rule 3 applies. Output: Plain muffin.
Order 4: Doesn't contain any of the specified ingredients in the rules, so rule 3 applies. Output: Plain muffin.

Q: 4 In the library, three books are displayed: "Mythology," "History," and "Science." You know the following:

- At least one of the books is fiction.
- "Mythology" is fiction if and only if "History" is non-fiction.
- "Science" is non-fiction if and only if "Mythology" is non-fiction.

Which of the following statements must be true?

A. All three books are fiction.

B. All three books are non-fiction.

C. Only "Mythology" is fiction.

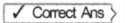 **D.** None of the above.

Explanation

Option (a): All three being fiction is possible if "History" and "Science" are both fiction, but it's not guaranteed by the given information. It depends on "Mythology" being fiction too.
Option (b): All three being non-fiction is also possible if "Mythology" and "History" are both non-fiction, but again, it's not guaranteed. "Science" could still be fiction based on the rules.
Option (c): Only "Mythology" being fiction is possible if "History" is non-fiction and "Science" follows suit, but as seen above, "Science" could still be fiction.

Therefore, none of the options guarantees a specific scenario and depends on the unknown status of "Mythology." This means the correct answer is: (d) None of the above.

Q: 5 A scientist conducts an experiment with three chemicals: X, Y, and Z. When mixed, they produce a specific color reaction:
- X and Y together produce no color.
- Y and Z together produce blue.
- X and Z together produce yellow.

Red and blue when mixed together create violet colour, and blue and yellow make green. What color will be produced when all three chemicals are mixed?

- **A.** Blue
- **B.** Yellow
- ✓ Correct Ans **C.** Green
- **D.** No color

Explanation

Since X and Y don't react alone, they won't interfere with the reaction between Y and Z, which produces blue. Similarly, X and Z's reaction (yellow) won't be inhibited by Y. Combining both reactions results in green.

Q: 6 A group of friends goes hiking. Each one takes same amout of time to hike. Sarah starts 30 minutes earlier than Emily, who starts 2 hours before John. If John takes 4 hours to complete the hike, how much earlier does Sarah finish compared to John?

- **A.** 1 hour 30 minutes
- **B.** 2 hours
- ✓ Correct Ans **C.** 2 hours 30 minutes
- **D.** 3 hours

Explanation

Emily -> 2 hrs earlier than John
Sarah -> 30 mins earlier than Emily.
Sarah finishes 2 hours 30 minutes earlier than John.

Q: 7 A museum displays ancient artefacts with labels in three languages: English, French, and Spanish. If an artefact has a label in both English and French, only then it has a Spanish label. However, some artefacts only have labels in one or two languages.
Which of the following statements must be true?

- ✓ Correct Ans **A.** Every artefact with a Spanish label also has an English label.
- **B.** Every artefact with a French label also has a Spanish label.
- **C.** Every artefact with an English label also has a French label.
- **D.** Every artefact with at least one label has a Spanish label.

Explanation

Spanish needs both English and French.

Q: 8 Sarah, Maya, and Daniel take turns watering three rose bushes, Rose A, Rose B, and Rose C. They follow a strict order: Sarah waters Rose A, then Maya waters Rose B, and finally, Daniel waters Rose C. After completing their rounds, they start again with Sarah watering Rose A.

If Sarah watered Rose A last Tuesday, who will water rose bush this coming Monday?

✓ Correct Ans **A.** Sarah
B. Maya
C. Daniel
D. Impossible to determine

Explanation

Since Sarah watered last Tuesday, the order continues. By Sunday, they will have completed two full cycles, placing Sarah to water Rose A.

Q: 9 In a local library each book has a unique code starting with a capital letter (F for fiction, N for nonfiction, B for biography), followed by three digits representing its specific category within the broader classification.

If a book with code "F235" is about historical fiction, what can you think as possibility about the following book codes?

A. N142 – It can be a book about science.
B. B589 – It can be a biography of a musician.
C. F001 – It can be the first book in the fiction category.
✓ Correct Ans **D.** All of the above are possible.

Explanation

All are possible based on first letter alone.

Q: 10 In the realm of enchantments, there are three potions: Elixir of Wisdom, Mystic Tonic, and Shrouded Brew. The Potion Master leaves a note:
• Potion A: "Only the wise can discern the Elixir of Wisdom."
• Potion B: "Those who trust their inner-self and their sixth sense, may find solace in the Mystic Tonic."
• Potion C: "For those who navigate the shadows, the Shrouded Brew awaits."

If Alex is known for relying on intuition, which potion should Alex choose?

A. Potion A
✓ Correct Ans **B.** Potion B
C. Potion C
D. Impossible to determine

Explanation

Since Alex relies on intuition, the choice aligns with the note on Potion B, mentioning trust in intuition for the Mystic Tonic.

Q: 11 In a distant galaxy, astronomers make a groundbreaking discovery of three celestial objects: Nebula X, Quasar Y, and Black Hole Z. Their observations reveal intriguing patterns:
- Pattern A: "Nebula X is always found near Quasar Y."
- Pattern B: "Black Hole Z exerts near gravitational influence over Nebula X."
- Pattern C: "Quasar Y emits energy pulses that Black Hole Z absorbs."

If astronomers locate Nebula X, which celestial object are they most likely to find nearby?

A. Quasar Y

B. Black Hole Z

C. Impossible to determine

✓ Correct Ans D. Both Quasar Y and Black Hole Z

Explanation

According to Pattern A, Nebula X is always found near Quasar Y. Additionally, Pattern B states that Black Hole Z exerts near gravitational influence over Nebula X. Hence, astronomers are likely to find both Quasar Y and Black Hole Z near Nebula X.

Q: 12 A group of time travelers embarks on a journey to three historical periods: Renaissance, Industrial Revolution, and Ancient Egypt. Their time machine follows specific guidelines:

- Guideline A: "The trip to Ancient Egypt precedes the visit to the Industrial Revolution."
- Guideline B: "If the Renaissance is the destination, the Industrial Revolution must follow."
- Guideline C: "The journey to Ancient Egypt is never the first destination."

Which historical period did the travellers visit first?

✓ Correct Ans A. Renaissance

B. Industrial Revolution

C. Ancient Egypt

D. Impossible to determine

Explanation

According to Guideline A, the trip to Ancient Egypt precedes the visit to the Industrial Revolution. Since they visited the Industrial Revolution after Ancient Egypt, the first destination must have been the Renaissance, as stated in Guideline B.

Q: 13 Intergalactic explorers—Stella, Victor, and Xavier—are on a quest for the legendary Galactic Treasure, navigating through cosmic challenges:
• Challenge A: "Stella often faces asteroid belt. If Stella encounters the asteroid belt, Victor must navigate through the gravitational anomaly immediately after."
• Challenge B: "Xavier, who is skilled in navigating gravitational anomalies, always precedes Stella in the treasure hunt."
• Challenge C: "If Victor successfully manoeuvres through the wormhole, Stella must encounter the asteroid belt immediately after."

If Xavier navigates through the wormhole, what possible challenge will Stella face next?

A. Gravitational anomaly
✓ Correct Ans B. Asteroid belt
C. Wormhole
D. Impossible to determine

Explanation

Since Xavier navigates through the wormhole, and according to Challenge B, Xavier always precedes Stella, it implies that Stella's next challenge will likely be facing the asteroid belt according to Challenge A.

Q: 14 Observe the following pattern of the numbers.

[8, 6, 2, 7]
[9, 3, 6, 2]
[7, 8, Q, 5]
[5, 7, 4, 3]

Find the value of **Q**.

A. 4
B. 2
✓ Correct Ans C. 3
D. 5

Explanation

The numbers are arranged: (First-term + Second term) ÷ Third term = Fourth term
So 7 + 8 ÷ Q =5; 15 ÷ Q = 5, Q = 3. Option C is correct.

Q: 15 **News Report**: "There is a massive earthquake in parts of Indonesia."

Jerry: "My friend Colin is in danger. He lives in Indonesia."

Which of the following shows the mistake in Jerry's reasoning?

    A. There was a reported fatality in the Indonesian incident.

    B. Earthquake is a devastating happening.

✓ Correct Ans **C.** The Earthquake did not affect every part of the country

    D. The magnitude of this earthquake is high.

Explanation

It has not been reported that this earthquake happened in the city where Colin lives. Option C is correct.

Q: 16 An examination's questions have the following mark allocation:
Q1→6 marks
Q2→4 marks
Q3→4 marks
Q4→4 marks
Q5→4 marks

If David got 12 marks for the paper, how many questions may he have answered correctly? (Students will be given either full or no mark for each question.)

    A. 4

    B. 2

✓ Correct Ans **C.** 3

    D. 1

Explanation

Apart from Question 1 with 6 marks, all other 4 questions have 4 marks each. For a student to score 12 marks that means he scored 0 in question 1 as 12 – 6 = 6, and the remaining 6 can't be scored from two questions. So he scored 3 questions with 4 marks each. Option C is correct.

Q: 17 Florence is sitting with 4 friends in a row on a bench. If there is an odd number of friends on either side, what is the position of Florence from the right?

    A. 2nd or 3rd

    B. 4th

✓ Correct Ans **C.** 2nd or 4th

    D. 3rd

Explanation

Two options:
5 people are sitting on the bench.
If 1 friend is on his right side, there are three to his left, he is in 4th position from the right
If 1 friend is on his left side, there are 3 friends on his right, he is in the second position from the right. Option C is correct.

Q: 18 Students are preparing for an excursion to the seaside. If Sam goes, Ray and Chris will not go, but Paul and Peter will go. If Pam goes, Stella and Basil go. Sam goes when Pam goes.

Pam is going to seashore. How many will go to the excursion among those mentioned?

    A. 5

    B. 4

    C. 3

✓ Correct Ans › **D. 6**

Explanation

The people who will go are Sam, Paul and Peter. Ray and Chris won't go because of Sam. Pam and Stella will also go with Basil. So Sam, Paul, Peter, Pam, Stella and Basil will go. Option D is correct.

Q: 19 Study the table of numbers below:

8	7	6	5	4
4	2	7	9	1
3	4	9	4	0

If column 3 and column 5 are interchanged and row 1 and row 3 are interchanged, what is the sum of the numbers in the 2nd column?

    A. 15

✓ Correct Ans › **B. 13**

    C. 5

    D. 22

Explanation

Interchanging columns 3 and 5, will not affect the sum of numbers in column 2. The sum in column 2 will be 7 + 2 + 4 = 13.
Then interchanging rows 1 and 3 will also not affect the sum of new column 2 which is 4 + 2 + 7 = 13. Option B is correct.

Q: 20 The following tables show the scores of four students in two Chemistry tests.

Test 1

Felix	71
George	72
Sally	69
Pam	53
Chris	61

Test 2

Felix	80
George	75
Sally	68
Pam	65
Chris	60

A prize is going to be given to the most improved student provided he/she has an average above 70 for both tests. Who should win the award?

A. George

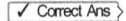 B. Felix

C. Sally

D. Pam

Explanation

The improvements over the two tests are:

- Felix: 71 to 80 = 9 marks. Average – 75.5
- George: 72 to 75 = 3 marks. Average – 73.5
- Sally: 69 to 78 = 8 marks. Average - 73.5
- Pam: 53 to 65 = 12 marks. Average - 59
- Chris: 61 to 60 = (-1) mark. Average – 60.5

Option B is correct.
The most improved is Pam (12) but her average is 59 below the mark to win the award. The next person is Felix = 9 marks with an average of 75.5. So, if he wins the award, Option B is correct.

Q: 21 John, the editor of a national newspaper needs to fill a vacancy for a Business Reporter in the newspaper. He needs someone who can write and investigate stories, is good with figures and economic terms and issues and also is readily available for frequent travels across the country.
He has shortlisted four applicants for this position:

i. Brandon: A married journalist, with five years of experience as a Business reporter. He has two young children and a wife, who is a nurse. Travel may be a problem for him.
ii. Ben, is a writer with about three years of experience as a travel blog writer. He is in his mid-20s and unmarried.
iii. Gray, has distinguished himself on the Business Desk of the local newspaper for the past three years, since he left the University four years ago. Loves to travel solo.
iv. Ray, who has a degree in Economics, has worked as a reporter for 4years on the Political Desk of another national newspaper.

Who should be offered the position?

A. Brandon
B. Ben
C. Gray ✓ Correct Ans
D. Ray

Explanation

Brandon looks experienced as a Business reporter but his family situation might preclude him from frequent travels as demanded by the position.
Ben is an experienced writer. But as a travel writer he might not be conversant with Economic and Business issues.
Gray, as a Business reporter has distinguished himself at the local newspaper. He is young and should be ready for the bigger space and can travel easily and frequently.
Ray, has an educational advantage and has been a reporter in a national newspaper. But his relative inexperience on the Business Desk might be to the disadvantage of the work required.
Gray should be employed: he has the experience and nothing to stop him from the frequent travels required.

Q: 22 Dauphin is a new hairdresser in town. Since she opened her new shop three weeks ago, she found out patronage has not been encouraging as most women still consider her an outsider in town.

If you are consulting for her business what will you advise her to do to save her business?

A. Start a new complimenting service to her hairdressing business.
B. Join the local women's groups and get involved in community services. ✓ Correct Ans
C. Start advertising her business on national newspaper.
D. Relocating her business out of town.

Explanation

She needs to make her neighbours and people around her feel comfortable with her and see her as one of them. This will encourage patronage and make people appreciate her expertise as a hairdresser.
She needs to interact with them, introduce her business and make them comfortable with her. Option B is the best strategy to win her new neighbours.

Q: 23 The final of the football match will be played on the 7th Sunday after the opening match. The opening match was on Friday, June 23rd.
When will the final match be played?

- A. September 3
- B. August 13
- ✓ Correct Ans C. August 6
- D. July 30

Explanation

The Opening match is on Friday, June 23. The next Sunday is June 25, 2 days later. The final will be 7 Sundays after the opening match, which is 2 + (6 X 7) days =44 days after June 23, the opening match day. This is August 6. Option C is correct.

Q: 24 If you go to bed late, you will wake up late next morning and miss your bus to work. If you miss your bus, you will get to work late. If you get to work late, $10 will be deducted from your wages that day.

If Paul gets to work late, which of the following is true?

- A. Paul left the house on time for his bus
- ✓ Correct Ans B. $10 was deducted from his day's wage
- C. Paul woke up early that day
- D. Paul slept early the night before.

Explanation

If Paul gets to work late $10 will be deducted from his day's wages. Other situations are inconclusive.

Q: 25 A posthumous event is when it is held in celebration of someone's death. Which event below is the best example of a posthumous event?

- A. Musical legend, Sir Bob was awarded with a Lifetime Achievement Award last year.
- B. The great Egyptian Writer could not receive his third Breaker Award personally because he was in self-imposed exile.
- ✓ Correct Ans C. The Award for the deceased boxer was accepted on his behalf by his first son, who himself is a boxing champion.
- D. The Award ceremony was cancelled because the main winners boycotted the ceremony because of their belief against racism.

Explanation

The only event whose major honoree is dead is the one for the deceased boxer.

Q: 26 From the following statements, answer the following question.
I. Raymond is 8 years old.
II. Barney likes Mathematics
III. Barney is the same age as Raymond.
IV. Barney and Raymond are taught by the same teachers
V. Raymond likes the English Language

Which of the statements helps us to know that Raymond and Barney are possibly classmates?

A. I, II, and III

✓ Correct Ans B. III and IV

C. I, II, V

D. All the above

Explanation

Classmates are usually the same age and are taught by the same teachers. So, Raymond are Barney are most likely classmates.

Q: 27 The Grand City Museum hosts three prestigious annual exhibitions: Art, History, and Science. Each exhibition has a unique ticket price: Art ($25), History ($30), and Science ($20). A special "Museum Explorer Pass" grants access to all three exhibitions for a discounted price. The discount covers cost of at least one exhibition. However, the pass is only available if purchased online in advance.
Sarah wants to visit all three exhibitions and is debating between buying individual tickets or the online pass.
What is the maximum price Sarah will pay if she decides for the Museum Explorer Pass online?

A. $50

✓ Correct Ans B. $55

C. $60

D. $65

Explanation

Since the combined cost of individual tickets is $75, the online pass must offer a discount to be cheaper. Analyzing the individual ticket prices, we see that the pass needs to offer a minimum discount of $20. This brings the maximum pass price to $75 - $20 = $55.

Q: 28 A museum exhibit showcases historical artefacts from different civilizations. Each artefact has a label displaying its origin, date, and material.
If a label reads "Egyptian, 1450 BC, Gold," which of the following artefacts cannot be described by this label?

A. A ceremonial mask

B. A funerary statue

✓ Correct Ans C. A decorative glass vase

D. A decorative pen

Explanation

Glass vase is made out of glass, not gold.

Q: 29 A group of friends, Emily, Michael, and Daniel, are planning a movie night. They each have different movie preferences: action, comedy, or drama. They share their thoughts:
- Emily: "I'm not in the mood for action tonight."
- Michael: "Drama isn't my thing, but I wouldn't mind a comedy."
- Daniel: "I'm open to action or comedy, but drama seems too slow."

Which movie genre will they most likely choose if they go together?

- A. Action
- ✓ Correct Ans B. Comedy
- C. Drama
- D. Any of the above

Explanation

Both Emily and Daniel eliminate action and drama, leaving comedy as the only option everyone is open to.

Q: 30 A city has three districts: East, West, and Central. Each district has a specific rule for crossing pedestrian bridges:
- East: You must be wearing a hat of a specific color.
- West: You must be carrying a specific type of flower.
- Central: You must be walking a dog.

You see Daniella crossing a bridge in the East district, not wearing a hat. Which of the following statements must be true?

- A. Daniella lives in the East district.
- B. Daniella is carrying a specific type of flower.
- C. Daniella is walking a dog.
- ✓ Correct Ans D. Daniella broke the rule for crossing the bridge in the East district.

Explanation

Daniella broke the rule for crossing the bridge in the East district.
Explanation: The information doesn't tell us where Daniella lives or if she carried flowers/walked a dog. However, we know she broke the East district's rule by not wearing the required hat.

Q: 31 Herminie and Ron send messages in a secret language. According to their language, "Be alert" is written as "Tr elaeb".

Ron sent "Tidnu of" to Herminie and what is the meaning of that?

- ✓ Correct Ans A. Found it
- B. Find it
- C. Found of
- D. None of the above

Explanation

Reverse the sentence without considering spaces (Bealert->Trelaeb)
Then place the spaces as in the original sentence (Be alert -> Tr elaeb)

Q: 32 If there is no electricity, Adam lights candles.
If Adam has invited someone to dinner, he lights candles.
Adam doesn't use candles otherwise.

There wasn't a power failure, but Adam was lighting candles. That likely means..

✓ Correct Ans **A.** Adam has invited someone to dinner.

B. Adam does not have invited someone to dinner.

C. Cannot decide about Adam having dinner with someone.

D. None of the above.

Explanation

By combining the given two statements, If there Is a power failure or Adam has invited someone to the dinner, then he lights candles.

Q: 33 Four traffic lights at an intersection follow a specific pattern:
• Light A turns green and stays green for 30 seconds.
• Light B turns green 15 seconds after Light A and stays green for 30 seconds.
• Light C turns green 15 seconds after Light B and stays green for 30 seconds.
• Light D turns green 15 seconds after Light C and stays green for 30 seconds.

If you arrive at the intersection when Light A is just turning green, what is the minimum amount of time you will have to wait for all lights to be green at the same time ?

A. 2 minutes

B. 1 minute 45 seconds

C. 1 minute 30 seconds

✓ Correct Ans **D.** Not possible

Explanation

As the gap between when light A turns green to when light D turns green is more than 30 secs, and light A is green for 30 secs only, it is not possible for all lights to be green at same time.

Q: 34 Two friends, Maya and Ethan, each flip a coin. Maya claims her coin landed heads, while Ethan claims his coin landed tails. However, you know that one of them is always lying and the other is always telling the truth.
What can you definitively conclude?

A. Maya's coin landed heads.

B. Ethan's coin landed tails.

C. At least one coin landed heads.

✓ Correct Ans **D.** None of the above.

Explanation

Since one friend always lies and the other always tells the truth, it is possible to have two heads or two tails.

Q: 35 A four-digit password is formed by using the digits 1, 2, 3, and 4. No digit can be repeated. You know that the first digit is odd, last digit is odd too, and the password is greater than 2000.
What could be the fourth digit?

✓ Correct Ans **A.** 1
B. 2
C. 3
D. 4

Explanation

The first digit is odd, so it can be 1 or 3. Since the password is greater than 2000, the first digit cannot be 1 (as 12xx, 13xx, etc. are all less than 2000). Therefore, the first digit must be 3, and the only remaining unused odd digit for the fourth position is 1.

Q: 36 Your body needs the vitamin B3 compound niacinamide to support the preservation of youthful-looking skin. Niacinamide, niacin and nicotinamide riboside are the three different forms of vitamin B3, however, niacinamide is the one that's most frequently seen in skincare products. Niacinamide makes your skin more moisturized and less sensitive by helping to lock in moisture and keep pollution or other potential irritants out. It has been demonstrated that niacinamide reduces inflammation, which can help lessen redness brought on by conditions including eczema, rosacea, and acne. Niacinamide can help keep your skin clear and smooth, which can reduce the appearance of pores and possibly cure dark spots, prevent skin cancer, and minimize fine lines and wrinkles.

If the information is true, which of the following must be false?

A. Niacinamide helps in keeping your skin moisturized.
B. Niacinamide can possibly treat dark spots.
C. Niacinamide can be used by those with acne problems.
✓ Correct Ans **D.** Niacinamide is not ideal to be used topically on skin.

Explanation

Based on the given information, niacinamide is the one that's most frequently seen in skincare products. Therefore, it can be topically used. There is no statement stating that it cannot be used topically on skin which makes option D to be a false statement.

Q: 37 During summer time, all tourists at Lagoon beach are reminded to bring and drink water or any kind of refreshment while spending their time under the sun. The owner of the Lagoon beach says that it can help prevent the cases of fatality at Lagoon beach during summer time.

Which of the following situations best strengthens the reminder of the owner of the Lagoon beach?

A. The owner of Lagoon beach wants to sell more of their refreshments especially during summer time.
✓ Correct Ans **B.** Cases of heat stroke that leads to death have happened at Lagoon beach during summer time over the past years.
C. The safety officer of Lagoon beach wants to help the local vendors at Lagoon beach during summer time.
D. The owner of Lagoon beach does not want any tourists at their beach during summer time.

Explanation
Tourists should bring fluid to avoid heat strokes.

Q: 38 Use the rule of the pattern and find the missing number which will replace the question mark.

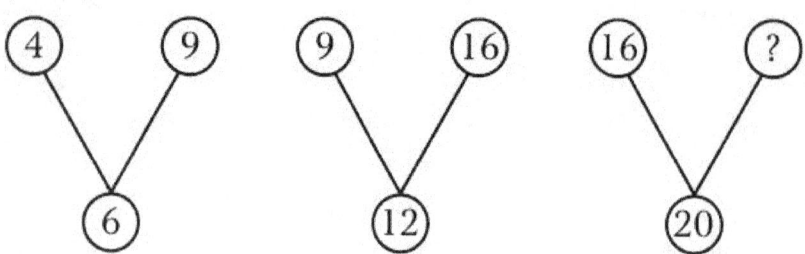

    A. 21

✓ Correct Ans B. 25

    C. 45

    D. 36

Explanation

(B): 4 x 9 = 36 → √36 = 6
9 x 16 = 144 → √144 = 12
Similarly, 16 x **X** → √16**X** = 20
On squaring :

16**X** = 400 $\therefore x = \dfrac{400}{16} = 25$

Q: 39 There are three separate large black boxes, and inside each large box there are two separate small red boxes; inside each of these small boxes, there is one smaller blue box.
How many boxes are there altogether?

    A. 9

    B. 12

✓ Correct Ans C. 15

    D. 18

Explanation

3 + 6 + 6 = 15

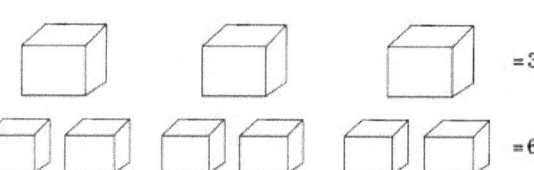

Q: 40 Compare the knowledge of persons X, Y, Z, A, B and C in relation to each other. X knows more than A. Y knows as much as B. Z know less than C. A knows more than Y. The best knowledgeable person amongst all is:

- A. C
- B. X
- C. A
- ✓ Correct Ans **D.** Cannot be determined

Explanation

X > A, Y = B
C > Z, X > A > Y = B
But we do not know the position of C and Z in relation to others.
So, (d) is the answer.

Preptive Prepare Better
Series L Reading Comprehension Trial Test

Answers and Solutions

Q: 1   In Extract B, what phrase encapsulates the central belief in the transformative power of kindness?

- A. "Unity in diversity"
- B. "Every cloud has a silver lining"
- ✓ Correct Ans C. "One good turn deserves another"
- D. "Actions speak louder than words"

Explanation

Extract B highlights the central belief in the transformative power of kindness through the phrase "One good turn deserves another," emphasizing the principle of reciprocity.

Q: 2   According to Extract A, what role does the ancient oak tree play in the community of Willowbrook?

- A. It serves as a gathering place for social events
- B. It provides shade for weary travelers
- ✓ Correct Ans C. It symbolizes the passage of time and wisdom
- D. It represents the town's commitment to environmental conservation

Explanation

Extract A suggests that the ancient oak tree symbolizes the passage of time and wisdom, serving as a significant cultural and historical landmark in the community of Willowbrook.

Q: 3   How does Liam respond to Mr. Thompson's tales in Extract A?

- A. He becomes bored and walks away
- ✓ Correct Ans B. He listens attentively and learns valuable lessons
- C. He interrupts Mr. Thompson repeatedly
- D. He falls asleep while Mr. Thompson speaks

Explanation

Extract A describes Liam as responding to Mr. Thompson's tales by listening attentively and learning valuable lessons about respect, empathy, and reciprocity.

Q: 4 What change occurs in Marcus's life as a result of Emily's gesture in Extract B?

- A. He decides to leave Metro City permanently
- B. He becomes more selfish and withdrawn
- ✓ Correct Ans C. He gains renewed inspiration in his art
- D. He loses interest in pursuing his passion

Explanation

Extract B mentions that Marcus gains renewed inspiration in his art as a result of Emily's gesture, leading him to create masterpieces that capture the spirit of the city.

Q: 5 What central message about reciprocity is conveyed through both extracts?

- A. Kindness should only be extended to those who deserve it
- B. Acts of generosity often lead to negative consequences
- ✓ Correct Ans C. Reciprocity enriches both the giver and the receiver
- D. It is important to expect something in return for acts of kindness

Explanation

Both extracts convey the central message that reciprocity enriches both the giver and the receiver, emphasizing the positive outcomes that arise from acts of kindness and generosity.

Q: 6 What distinguishes the Mona Lisa and David as discussed in Extract A?

- A. Their simplicity and minimalist design
- ✓ Correct Ans B. The intricate details and symbolism infused by the artists
- C. Their significant monetary value in the art market
- D. The use of vibrant colors and bold brushstrokes

Explanation

Extract A highlights the meticulous attention to detail and profound symbolism incorporated by artists like Leonardo da Vinci and Michelangelo in creating masterpieces like the Mona Lisa and David.

Q: 7 What aspect of Shakespeare's Hamlet is highlighted in Extract A?

- ✓ Correct Ans A. Its role as a timeless exploration of existential themes
- B. Its popularity as the most widely performed play
- C. Its portrayal of traditional values and customs
- D. Its use of innovative theatrical techniques

Explanation

Extract A focuses on Shakespeare's Hamlet as a profound exploration of existential angst, moral ambiguity, and the complexities of the human condition, highlighting its enduring relevance and impact.

Q: 8 According to Extract A, what defines the greatest artwork?

- A. Its monetary value in the global art market
- B. The technical proficiency of the artist
- ✓ Correct Ans **C. Its emotional resonance and enduring impact**
- D. The popularity and fame of the artist

Explanation

Extract A suggests that the greatest artwork is defined by its emotional resonance and enduring impact on individuals and society, transcending mere technical proficiency or monetary value.

Q: 9 How does Extract B emphasize the role of art in times of adversity?

- A. By discussing its monetary value as an investment
- B. By analyzing its influence on fashion and design trends
- ✓ Correct Ans **C. By highlighting its capacity to offer solace and hope**
- D. By comparing it to traditional cultural artefacts

Explanation

Extract B underscores the role of art in providing solace, comfort, and hope in times of adversity, emphasizing its ability to uplift the human spirit and inspire resilience.

Q: 10 How does Extract B portray the value of indigenous crafts and street art?

- A. As commodities with high market demand
- ✓ Correct Ans **B. As reflections of cultural identity and heritage**
- C. As symbols of elitism and exclusivity
- D. As products of mainstream art institutions

Explanation

Extract B depicts indigenous crafts and street art as expressions of cultural identity and heritage, emphasizing their significance beyond their commercial value.

Q: 11 According to Extract B, how does art contribute to societal dialogue and activism?

- A. By conforming to established cultural norms and values
- B. By serving as a status symbol for the elite class
- C. By encouraging passive consumption and enjoyment
- ✓ Correct Ans **D.** By provoking thought, sparking dialogue, and galvanizing movements for justice

Explanation

Extract B suggests that art contributes to societal dialogue and activism by provoking thought, sparking dialogue, and galvanizing movements for justice, rather than conforming to established norms or serving as a status symbol.

Q: 12 Which extract emphasizes the artistry of storytelling, comparing it to a ballet performance?

- ✓ Correct Ans **A.** Extract A
- B. Extract B
- C. Extract C
- D. Extract D

Explanation

The correct answer is (a) Extract A. It describes narrative writing as an artistry that breathes life into characters and orchestrates plots as intricate as a ballet performance.

Q: 13 In which extract is writing portrayed as a canvas for linguistic experimentation and emotional exploration?

- A. Extract A
- B. Extract B
- C. Extract C
- ✓ Correct Ans **D.** Extract D

Explanation

The correct answer is (d) Extract D. Creative writing is depicted as an expansive universe for artistic exploration and the birth of new worlds through language.

Q: 14 Which extract likens narrative writing to a symphony that resonates in readers' hearts and minds?

[✓ Correct Ans] **A.** Extract A
B. Extract B
C. Extract C
D. Extract D

Explanation

The correct answer is (a) Extract A. It describes narrative writing as creating a symphony that resonates in readers' hearts and minds.

Q: 15 In which extract is writing described as an influential force commanding attention and igniting action?

A. Extract A
B. Extract B
[✓ Correct Ans] **C.** Extract C
D. Extract D

Explanation

The correct answer is (c) Extract C. Persuasive writing is portrayed as an influential force wielding reason, emotion, and eloquence to sway opinions and ignite action.

Q: 16 Which extract portrays writing as a pursuit to elucidate complex concepts and empower readers with insights?

A. Extract A
[✓ Correct Ans] **B.** Extract B
C. Extract C
D. Extract D

Explanation

The correct answer is (b) Extract B. Expository writing is described as a beacon of informative discourse, elucidating complex concepts and empowering readers with insights.

Q: 17 Which extract highlights a form of writing that focuses on unravelling complex concepts and providing clear insights to its readers?

A. Extract A
[✓ Correct Ans] **B.** Extract B
C. Extract C
D. Extract D

Explanation

Extract B. Expository writing is described as shedding light on complex subjects and providing clear insights.

Q: 18 Which one fits in (1) ?

- A. Sentence A
- B. Sentence B
- ✓ Correct Ans C. Sentence G
- D. Sentence H

Explanation

The correct answer is "c" because the starting line should be about the description about the traditional tourism at first and then move forward to new ways of tourism

Q: 19 Which one fits in (2) ?

- ✓ Correct Ans A. Sentence A
- B. Sentence B
- C. Sentence C
- D. Sentence D

Explanation

The correct answer is "a" because the prior line states about responsibility of conserving. For that guides are important

Q: 20 Which one fits in (3) ?

- A. Sentence A
- B. Sentence D
- ✓ Correct Ans C. Sentence B
- D. Sentence E

Explanation

The correct answer is "c" because the next part of the paragraph describes about various things that can be done through eco-tourism.

Q: 21 Which one fits in (4) ?

- A. Sentence A
- B. Sentence B
- C. Sentence F
- ✓ Correct Ans D. Sentence E

Explanation

The last line of this paragraph gives a hint that Despite its significant advantages, ecotourism is not without challenges. Thus, challenges should come to this place.

Q: 22 Which one fits in (5) ?

- A. Sentence F
- B. Sentence B
- ✓ Correct Ans **C. Sentence C**
- D. Sentence D

Explanation

The first line depicts the future of the tourism. So that correct answer should support that fac.

Q: 23 Which one fits in (6) ?

- A. Sentence A
- B. Sentence B
- C. Sentence C
- ✓ Correct Ans **D. Sentence D**

Explanation

The correct answer's outcome is stated in the next line.

Q: 24 Which one fits in (7) ?

- A. Sentence A
- B. Sentence G
- ✓ Correct Ans **C. Sentence H**
- D. Sentence C

Explanation

The correct answer is "c" because the sentence should be an end note to the whole essay.

Q: 25 What is the tone of the poem "The Listeners"?

- A. Joyful
- ✓ Correct Ans **B. Mysterious**
- C. Angry
- D. Sad

Explanation

The tone of the poem is mysterious as it revolves around the unanswered question and the eerie atmosphere of the empty house.

Q: 26 What does the poet convey through the line "That I kept my word"?

☑ Correct Ans **A.** The Traveller kept his promise.
B. The Traveller broke his promise.
C. The Traveller was lying.
D. The Traveller was angry.

Explanation

The line conveys that the Traveller fulfilled his promise by visiting the house, but no one answered.

Q: 27 What is the deeper meaning of the poem "The Listeners"?

A. The importance of communication
B. The significance of keeping promises
☑ Correct Ans **C.** The mystery of existence and unanswered questions
D. The loneliness of the Traveller

Explanation

The deeper meaning of the poem revolves around the mystery of existence and the unanswered questions that surround it, as represented by the listeners in the poem.

Q: 28 What does the poet mean by "leave kind words as mementoes"?

A. Kind words are important
B. Kind words are essential
☑ Correct Ans **C.** Kind words can be remembered by others
D. None of the above

Explanation

The poet means that we should leave behind kind words that can be cherished and remembered by others.

Q: 29 What is the significance of the phrase "watch the noon-time hour arrive in gold and tinsel dressed"?

☑ Correct Ans **A.** Kind words can turn ordinary things into something beautiful
B. Kind words are useless
C. Kind words are essential
D. None of the above

Explanation

The phrase suggests that speaking kindly can transform an ordinary moment into something beautiful and magical.

Q: 30 What does the poet mean by "smooth the flaxen hair"?

✓ Correct Ans **A.** Running hand through a child's hair

B. Running hands through a boy's hair

C. Running hands through an adult's hair

D. None of the above

Explanation

The poet is referring to gently running one's hand through a child's hair.

Preptive Prepare Better
Series L Mathematical Reasoning Trial Test

Answers and Solutions

Q: 1 The **average** mass of Alex, Ben and Charles is 49 kg.
Alex is 9 kg heavier than Ben and 6 kg heavier than Charles.

What is Charles' mass?

- A. 44 kg
- B. 45 kg
- ✓ Correct Ans → C. 48 kg
- D. 50 kg

Explanation

Alex is **a**, Ben is **b** and Charles is **c**.

$a = b + 9$; $b = a - 9$
$a = c + 6$; $c = a - 6$

Now, $a + b + c = 49 \times 3 = 147$

$a + (a-9) + (a-6) = 147$; $a \times 3 = 147 + 15 = 162$
$a = 54$
$c = a - 6 = 54 - 6 = 48$ kg

Q: 2 Pablo can choose ham, turkey, tomato, or cheese on his sandwich.
If he chooses two different toppings, how many different sandwiches can he make, assuming order of toppings matters?

- A. 6
- ✓ Correct Ans → B. 12
- C. 24
- D. 8

Explanation

First topping = 4 choices
Second topping = 3 choices, as first and second topping needs to be different.

Total choices = 4×3 = 12 choices

Q: 3 There were an equal number of boys and girls in the Art club at first. Then, another 17 boys joined the club and another 7 girls left the club.
There are four times as many boys as girls in the club now.
How many girls were there at first?

✓ Correct Ans **A.** 15
B. 25
C. 24
D. 28

Explanation

At first there are 'n' boys and 'n' girls.

(n-7) x 4 = n + 17; nx3 = 17 + 28 = 45; n = 15

Q: 4 Mrs Lim bought 4 kg of strawberries.
Li Ming ate $\frac{1}{4}$ of it and Ming Hui ate $\frac{3}{5}$ kg.
How many kilograms of strawberries were left?

A. 2 3/5 kg
✓ Correct Ans **B.** 2 2/5 kg
C. 2 1/2 kg
D. 1 2/5 kg

Explanation

Li Ming → ¼ x 4kg = 1kg
Ming Hui → 3/5 kg
Total eaten → 1 kg + 3/5 kg = 1 3/5 kg
Left → 4 kg - 1 3/5 kg = 2 2/5 kg

Q: 5 Ema had 4 times as many beads as Ben at first.
After Ema gave 60 beads to Ben, both of them had the same number of beads.

How many beads did each of them have at first?

- A. 180, 30
- B. 160, 20
- C. 150, 40
- ✓ Correct Ans **D.** 160, 40

Explanation

At start : Ema = 4n, Ben = n
At end : Ema = 4n - 60, Ben = n + 60

4n - 60 = n + 60
3n = 120; n = 40
So at first, Ema = 160, Ben = 40

Q: 6 A bakery shop baked 10.05 kg of cookies. Some of the cookies were packed into small packets of 250 g each and the rest was packed into big packets of 600 g each. In the end, the number of big packets was 4 more than the number of small packets.
How many big packets of cookies were packed?

- ✓ Correct Ans **A.** 13
- B. 18
- C. 23
- D. 33

Explanation

There are 'n' small packets.

Small: n x 250 = 250n
Big: (n + 4) x 600 = 600n + 2400

10.05kg = 10050g
250n + 600n + 2400 = 10050
850n = 7650; n = 9; n+4 = 13

Q: 7 The table shows the length of four ribbons.

Ribbon	A	B	C	D
Length (cm)	11.3	6.5	?	?

The average length of the 4 ribbons is 9.6 cm.
Write down one possible set of lengths for Ribbon C and Ribbon D.

  A. 10 cm, 11.6 cm

  ✓ Correct Ans B. 11 cm, 9.6 cm

  C. 8.2 cm, 12.9 cm

  D. None of these

Explanation

9.6 X 4 = 38.4
11.3 + 6.5 = 17.8
Sum of C and D = 38.4 - 17.8 = 20.6 cm
Answer is any 2 values that add up to 20.6 cm.

Q: 8 In July, Raees, Leon and James saved a total of $1200. In August, Raees doubled his savings, Leon decreased his savings by $160 and James increased his savings by $110. Their savings were the same in August.
What was Raees' savings in August?

  A. $360

  B. $220

  C. $720

  ✓ Correct Ans D. $460

Explanation

In August : R = n, L = n, J = n
In July : R = $n/2$, L = n+160, J = n - 110
$n/2$ + n+160 + n-110 = 1200
$5/2$ x n + 50 = 1200
$5/2$ x n = 1150
n = 460

Q: 9 ABCD is a trapezium. ACD is an isosceles triangle where AD = DC. ∠ABC = 74° and ∠ACB = 86°. Find ∠ADC.

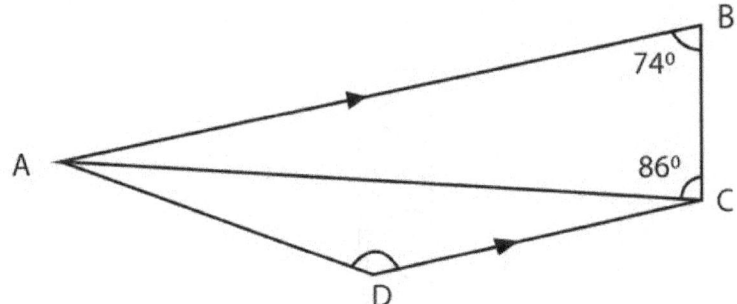

✓ Correct Ans **A.** 140°

B. 100°

C. 120°

D. 160°

Explanation

74 + 86 = 160
180 - 160 = 20
180 - 20 - 20 = 140°

Q: 10 The diagram below shows a bar chart.

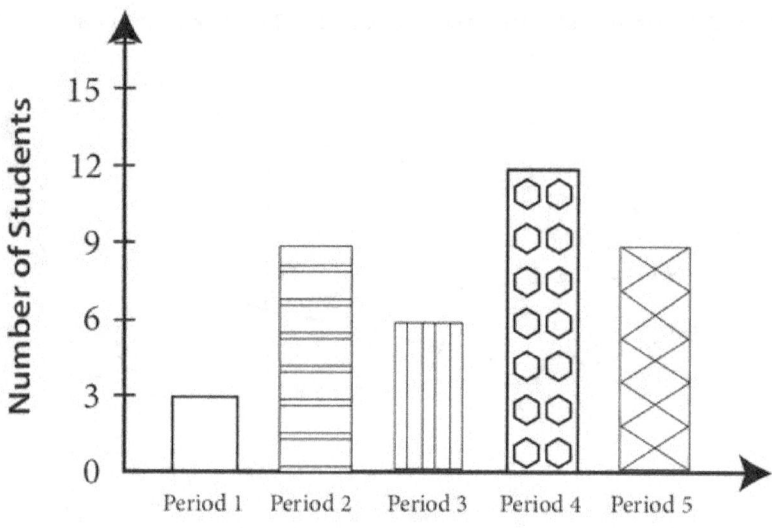

Class Categories

Which of the following is a false statement?

- A. The number of students in all of the class categories is 39
- B. The difference in the number of students between Period 4 and period 5 is 3
- C. The total number of students in periods 3, 4 and 1 is 21
- ✓ Correct Ans D. The difference in the number of students between the highest and the lowest is 10

Explanation

A. The number of students in all of the class categories is 39. Correct. 3+9+6+12+9=39
B. The difference in the number of students between Period 4 and period 5 is 3. Correct. 12-9=3
C. The total number of students in periods 3, 4 and 1 is 21. Correct. 6+12+3=21
D. The difference in the number of students between the highest and the lowest is 10. Wrong. 12-3=9

Q: 11 Kelly bought some stickers. He gave $4/6$ of them to his brother and $2/5$ of the remainder to his cousin. After that, Kelly had 66 stickers left. How many stickers did Kelly buy?

- ✓ Correct Ans A. 330
- B. 210
- C. 120
- D. 190

Explanation

Brother = $2/3$
Remaining = $1/3$
Cousin gets = $2/15$

Left = 1 - $2/3$ - $2/15$
= $15/15$ - $10/15$ - $2/15$ = $3/15$ = $1/5$

$1/5$ is 66. Whole is 330.

Q: 12 The figure below shows a rectangle ABCD. The ratio of AB to BC is 3 : 2. AB is 90 cm. What is the perimeter of rectangle ABCD?

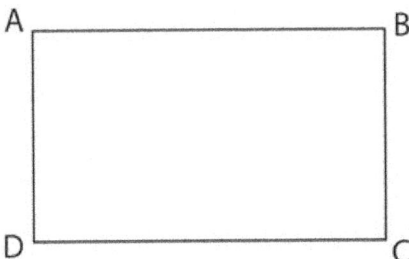

- A. 225 cm
- B. 740 cm
- ✓ Correct Ans C. 300 cm
- D. 150 cm

Explanation

3u = 90
1u = 90 ÷ 3 = 30
Perimeter of ABCD, 10u = 10 x 30 = 300cm

Q: 13 The pie chart shows the results of a survey conducted to identify the favorite game of some students.

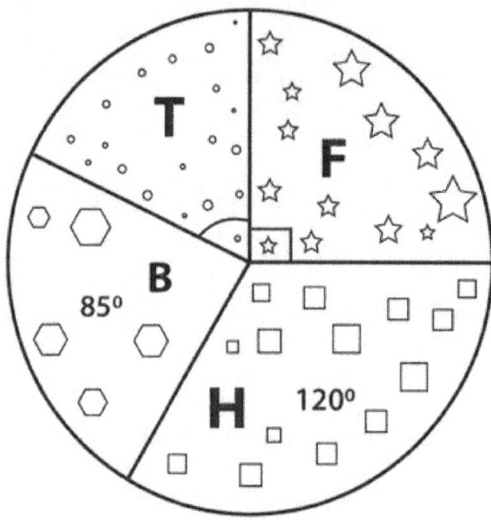

F Football H Hockey B Badminton T Tennis

How many students like Tennis if the total number of students is 360?

    A. 100

    B. 90

    C. 55

✓ Correct Ans **D.** 65

Explanation

Tennis = 360 − 85 − 120 − 90 = 65°
65/360 x 360 = 65 students

Q: 14 The given figure shows a party cap.

What is the cross-section obtained when a horizontal cut parallel to the base is given to the cap?

✓ Correct Ans **A.** Circle

    B. Cone

    C. Triangle

    D. Rectangle

Explanation

Upon horizontal cut parallel to the base, a circle is obtained.

Q: 15 Yoga classes were held for the following duration in a particular week.

Day	Duration of exercise (min)
Sunday	40
Monday	50
Tuesday	60
Wednesday	x
Thursday	30
Friday	90
Saturday	90

If the average duration of yoga class in the week is 60 minutes, what was the duration of the yoga class on Wednesday?

✓ Correct Ans **A.** 60 min
B. 30 min
C. 55 min
D. 65 min

Explanation

Average duration = (40 + 50 + 60 + n + 30 + 90 + 90) / 7 = 60
150 + n + 210 = 60 x 7
n = 420 − 360 = 60

Q: 16 Four friends - Dan, Fran, Ian, and Jan - went strawberry picking last Saturday.

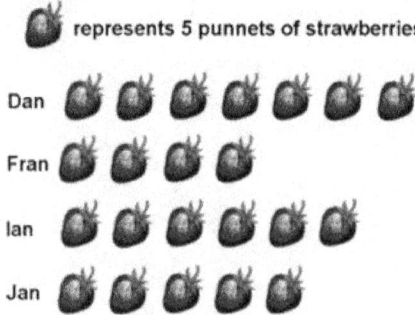

The pictograph shows the number of punnets of strawberries each of them picked. If this information is displayed instead as a pie chart, what would be the angle (to the nearest degree) of the sector of the pie chart representing the strawberries picked by Jan?

A. 25°
B. 50°
C. 65°
✓ Correct Ans D. 82°

Explanation

Dan picked 7 × 5 = 35 punnets
Fran picked 4 × 5 = 20 punnets
Ian picked 6 × 5 = 30 punnets
Jan picked 5 × 5 = 25 punnets

The total number of punnets picked = 35 + 20 + 30 + 25 = 110
Therefore, the fraction of the strawberries picked by Jan = 25/110

So the angle of the sector of the pie chart representing ice creams sold on Monday
= 25/110 × 360°
= 81.81...° = 82° to the nearest degree.

Q: 17 What is the y-intercept of the following line?

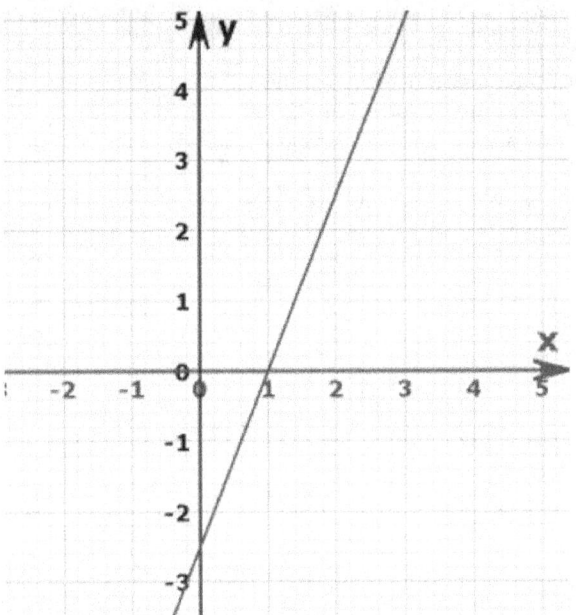

✓ Correct Ans A. (0, -2.5)
B. (0, -1)
C. (0, 1)
D. (0, 2.5)

Explanation

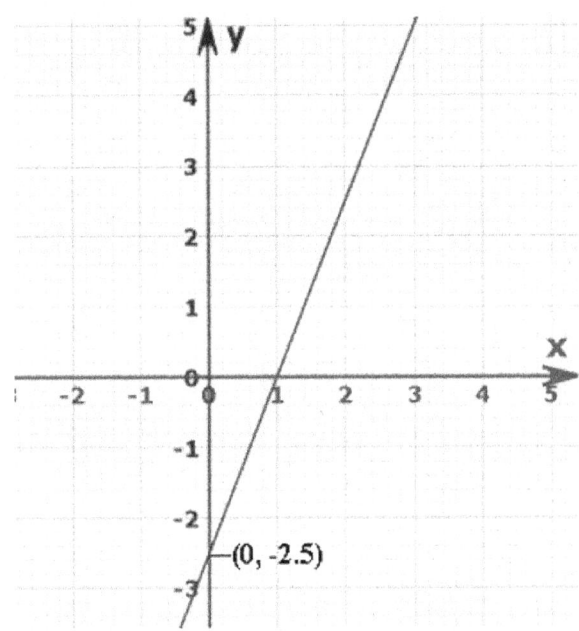

The y-intercept is the point where the line crosses the y-axis = (0, -2.5)

Q: 18 Four classes 5A, 5B, 5C and 5D donated money at a Charity event.
5A, 5B and 5C donated a total of $380. 5C and 5D donated a total of $208.
The ratio of the amount donated by 5C to the total amount donated by the four classes is 1 : 6.
What is the amount donated by 5C?

- A. $44
- B. $64
- C. $72
- ✓ Correct Ans D. $84

Explanation

A + B + C + D = 6 parts; C = 1 part
So, A + B + 2C + D = 6 + 1 = 7 parts

7 parts = 380 + 208 = 588; 1 part = 84

Q: 19 Wesley and Xavier have 217 marbles altogether. Xavier and Yixian have 105 marbles altogether. Wesley has 3 times as many marbles as Yixian.
How many marbles does Xavier have?

- A. 59
- B. 112
- C. 56
- ✓ Correct Ans D. 49

Explanation

W + X = 217 .. (1)
X + Y = 105 .. (2)
W = Yx3 .. (3)

(1) - (2) ;
W - Y = 217 - 105 = 112 .. (4)

From (3) and (4) :
Yx3 - Y = 112; Y = 56

From (2) : X = 105 - Y = 105 - 56 = 49

Q: 20 The working hours are 9:00 am to 5:00 pm, and in between 60 minutes are spent on lunch. Find the ratio of time spent on work to the time spent on lunch.

- A. 8:30
- ✓ Correct Ans **B. 7:1**
- C. 7:5
- D. 12:7

Explanation

Office hours = 9.00 am to 5.00 pm = 8.0 hours = 480 minutes
480 – 60 minutes for lunch = 420 mins
420: 60 = 7 : 1

Q: 21 Ms Nami runs the school cafeteria. On April, she had to order **15 cases of caramel pudding twice** due to the high demanding from her students. **Each contains p** of caramel pudding.

Choose the expressions that represent how many caramel puddings Ms Nami ordered on April.

- ✓ Correct Ans **A. 30p**
- B. 2p + 15p
- C. p(2 + 15)
- D. 3(15p)

Explanation

Multiply how many puddings Ms Nami ordered in April by how many pudding are in
Ms Nami ordered pudding twice, 2, times 15.
Therefore 2 x15 x p = 30p

Q: 22   A candy dispenser put various numbers of orange candies into bags.

Orange candies per bag

Stem	Leaf
2	4
3	0 3 5
4	4 6
5	5 6 6 7
6	4 5 6
7	3 4 7 9
8	2 5 5 5 6 8 8
9	0

How many bags had at least 20 orange candies?

A. 30

B. 29

C. 27

✓ Correct Ans D. 25

Explanation

25 leaves were counted, 25 bags had at least 20 orange candies.

Q: 23 The table shows the number of beads in a container.

Color	Number of Beads
White	180
Red	160
Blue	$\frac{2}{5}$ of the number of white beads

Daisy uses $1/4$ white beads and 20% of red beads to embroider her dress.

Count the number of beads left in the container.

- A. 275
- B. 298
- C. 335 ✓ Correct Ans
- D. 333

Explanation

Blue = $2/5$ x 180 = 72

White used = 1/4 x 180 = 45, balance 135

Red used = 160 x 20% = 32, balance 128
Remaining: 72 + 135 + 128 = 335

Q: 24 Which of the following statements are correct?

X. $2/3 + 2/3$ is more than $1 \frac{1}{2}$
Y. $1 - 2/6$ is less than $1/6$
Z. $1/5$ is more than $1/12$

- A. x and y only
- B. y and z only
- C. y only
- ✓ Correct Ans D. z only

Explanation

X. $2/3 + 2/3 = 4/3 = 8/6$
$1 \frac{1}{2} = 3/2 = 9/6$
X is incorrect

Y. $6/6 - 2/6 = 4/6$ is more than $1/6$
Y is incorrect.

Z. $1/5 = 12/60$
$1/12 = 5/60$
Z is correct

Q: 25 Emmy is redecorating her house. She decides to paint the half of the wall white. On top of the white paint, she paints orange, extending up to $1/5$ of the unpainted wall. As for the rest of the wall, she settles for yellow.

If the wall was divided into 10 equal parts, what would be the difference between the part painted orange only and yellow?

- A. 1 part
- B. 4 parts
- ✓ Correct Ans C. 3 parts
- D. 5 parts

Explanation

Orange only = $1/5 \times 1/2 = 1/10$, so $1/10 \times 10 = 1$-part painted orange.
What was painted yellow was $4/5 \times 1/2 = 4/10 = 2/5$. So, $2/5 \times 10 = 4$; 4 parts were painted yellow.
The difference is $4 - 1 = 3$ parts.

Q: 26 These are measure of time spent a day in various rooms by Joe. Data is for 2 days. A day has 24 hrs.

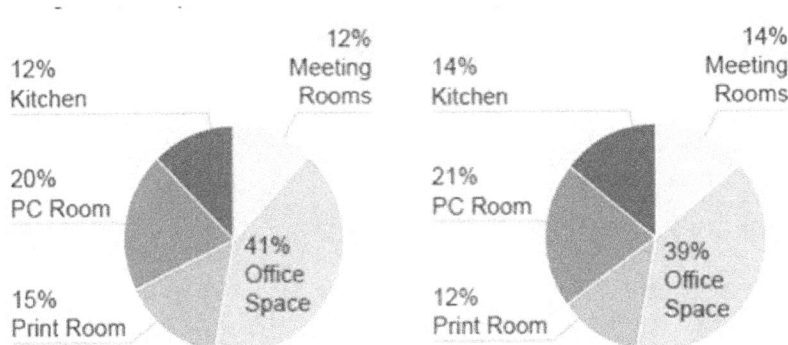

What's the difference between time spent in the kitchen between the first and second days, in hours ?

✓ Correct Ans **A.** 0.48 hrs

B. 0.64 hrs

C. 0.12 hrs

D. 0.24 hrs

Explanation

2% of 24 hrs
= 24 x $^2/_{100}$ = $^{48}/_{100}$ = 0.48

Q: 27 Kevin baked some pies.

He sold $\frac{1}{3}$ of the pies in the morning. He sold $\frac{1}{4}$ of the remainder in the afternoon. He had 156 pies left. How many pies did Kevin bake?

- A. 302
- B. 212
- ✓ Correct Ans C. 312
- D. 324

Explanation

Morning = $1/3$
Remainder = $2/3$
Afternoon = $2/3 \times 1/4 = 1/6$
Final remainder = $2/3 - 1/6 = 1/2$

$1/2$ is 156. Whole is 312.

Q: 28

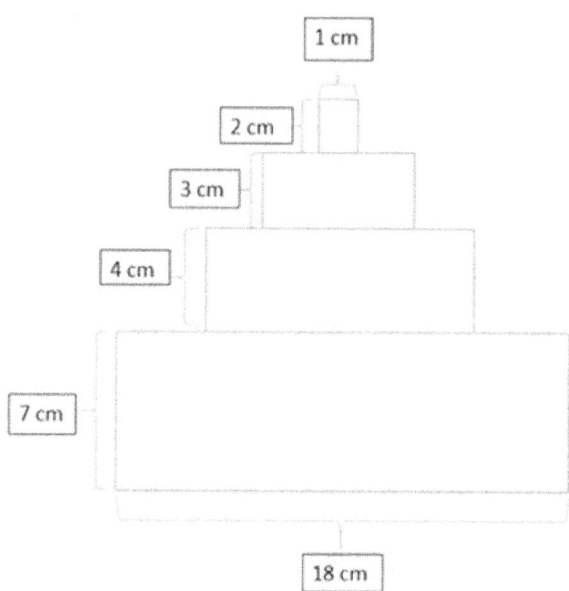

Find the perimeter of the above figure.

- ✓ Correct Ans A. 68 cm
- B. 70 cm
- C. 72 cm
- D. 74 cm

Explanation

Small horizontal lines give the sum of 18 cm.
Perimeter = 18+18+7+4+3+2+2+3+4+7 = 68 cm

Q: 29 Below shows the prices of some items at a bookshop.

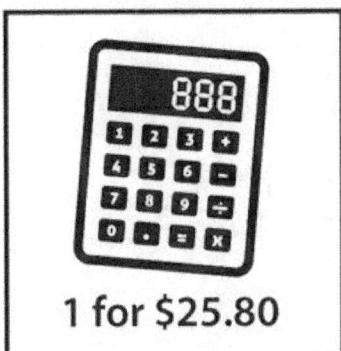
1 for $25.80

4 for $5.40

Kenny bought 2 calculators and 16 notebooks for $60.30. There was a **discount given on the calculators only**. What was the percentage discount of the calculators?

✓ Correct Ans **A.** 25%

B. 50%

C. 75%

D. 40%

Explanation

16 notebooks = 4 x 5.40 = 21.60
2 calculators = 60.30 - 21.60 = 38.70
2 calculators original = 25.80 x 2 = 51.60
Diff = 51.60 - 38.70 = 12.90
$$\frac{12.9}{51.6} \times 100\% = 25\%$$

Q: 30 The average of three 2-digit numbers is 45. One of the numbers is 70. What is the largest difference between the other two numbers?

A. 55

✓ Correct Ans **B.** 45

C. 65

D. 25

Explanation

Sum of 3 numbers → 45 × 3 = 135
Sum of other 2 numbers → 135 – 70 = 65
Smallest number possible → 10
Larger number → 65 – 10 = 55
Largest difference → 55 – 10 = 45

Q: 31 In the figure below, find the sum of ∠a, ∠b, ∠c, ∠d, ∠e, and ∠f.

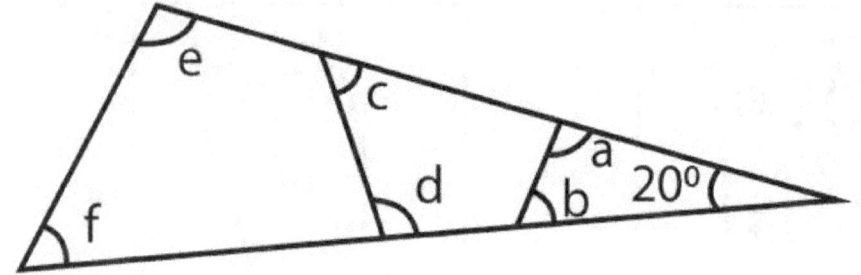

- A. 160°
- B. 240°
- C. 360°
- ✓ Correct Ans **D.** 480°

Explanation

$a + b = 180° - 20° = 160°$
$c + d = 180° - 20° = 160°$
$e + f = 180° - 20° = 160°$
$\angle a + \angle b + \angle c + \angle d + \angle e + \angle f = 160° \times 3 = 480°$

Q: 32

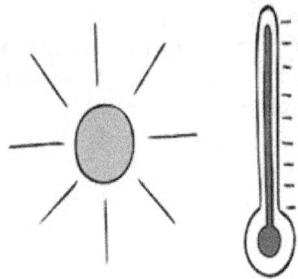

The average temperature of a town for the first 4 days of a month was 58°. The average temperature of the second, third, fourth and fifth day of this week was 60°.

If the ratio of the temperature of the first day and fifth day of the week is 7:8, then what was the temperature of the fifth day?

- A. 62°
- B. 52°
- ✓ Correct Ans C. 64°
- D. 72°

Explanation

According to question we can say that
M + T + W + Th = 58x4 = 232 .... (1)
T + W + Th + F = 60x4 = 240 ....(2)

Eq. (2)- (1) :
F- M = 8
Let's assume M = 7x and F = 8x
Then 8x – 7x = 8, x = 8
F = 8x = 64°
Hence temperature of the fifth day = 64°

Q: 33 Alex and Belle had $88. Alex saved $1/3$ of his money while Belle spent $4/5$ of her money. Given that Alex and Belle spent the same amount of money, how much did Alex save?

- ✓ Correct Ans A. $16
- B. $12
- C. $14
- D. $18

Explanation

	Start	Spent	Saved
Alex	n	$2/3$ n	$1/3$ n
Belle	88 - n	$4/5$ (88 - n)	$1/5$ (88 - n)

$2/3$ n = $4/5$ (88 - n)
10 n = 12x(88 - n)
10n = 12x88 - 12n
22n = 12x88; n = 12x4 = 48
Alex saved = $1/3$ n = 16

Q: 34 Mei placed some potted plants in a row from one end to the other end of the corridor. They were placed at an equal distance from one another. The distance between the 9th and the 13th potted plant was 24 m. If the length of the entire stretch of corridor was 78 m, find the total number of plants that were placed altogether.

✓ Correct Ans A. 14m
B. 16m
C. 15m
D. 12m

Explanation

13 − 9 = 4 (Number of gaps from 9th to 13th potted plants)
24 m ÷ 4 = 6 m
78 m ÷ 6 m = 13 (Total number of gaps)
13 + 1 = 14
There were 14 potted plants placed altogether.

Q: 35 Nurul and Mei Shan had an equal number of sweets at first. After Nurul gave away 192 sweets and Mei Shan gave away 24 sweets, Mei Shan had 4 times as many sweets as Nurul.

How many sweets did each girl have at first?

A. 244
B. 225
✓ Correct Ans C. 248
D. 256

Explanation

At start : Nurul = n, Mei Shan = n
At end : Nurul = n - 192, Mei Shan = n - 24

$n - 24 = (n - 192) \times 4 = 4n - 768$
$3n = 768 - 24 = 744; n = 248$

Preptive Prepare Better
Series L Thinking Skill Trial Test

Answers and Solutions

Q: 1 **Paul's mum:** If you pass the Mathematics test, you can go on an excursion to the seaside.

Paul didn't go on an excursion.

Which of the following is the most correct conclusion?

  A. Paul didn't pass the Mathematics test
  B. Paul passed the Mathematics test
  C. Paul passed the Mathematics test but decided not to go for the excursion
  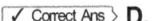 D. Can't decide

Explanation

Paul's mum permitted him to go for the excursion if he passed the test. But Paul can decide whether he is going on the excursion or not after passing the test. So, just knowing Paul didn't go on the excursion does not mean we should conclude that he passed or failed the test. Option D is correct.

Q: 2 In a video game, each game won gives the players the same number of points which gradually leads to a payout level. There are 4 payout levels in the game.

Level 1: The number of games won is less than that of Level 2.
Level 2: The number of games won is less than Level 3
Level 3: The number of games won is less than Level 4
Level 4: The number of games won is more than Level 3

Based on the information, can you determine ...

  A. Number of games won in Level 2 is double those in 1
  B. The number of games won in Level 3 is double the same in Level 1
   C. Inadequate information to determine the number of games in each level
  D. The number of games won in Level 4 is double those of Level 2

Explanation

There are no specific numbers of games in each level but the numbers were specified through inequalities. So, we have relative values of the number of games in each level.
While options A, B and D may be possible we do not have the exact relationship between the number of games in levels. So, to determine the relationship of the number of games in each level, we need more information. Option C is correct.

Q: 3 Jack Carlos has just been appointed the interim coach of Tradford United. He will be in charge for the rest of the season for the next 12 matches.
If he manages to qualify the club for Intercontinental competition by the end of the season, he will be confirmed as the chief coach for next season. Otherwise, a new coach will be appointed for next season.

Pedro: "I came to know a new coach won't be appointed for Tradford United next season. That means Carlos did a good job as the interim coach."
Ben: "Carlos can only serve as the interim coach for the rest of the season until a new coach (not Carlos) is appointed for the team next season, no matter what is team's result."

Which of the following reasoning is true?

 **A.** Pedro Only
B. Ben Only
C. Both Pedro and Ben
D. Neither Pedro nor Ben

Explanation

Both statements mention the conditions that might make Pedro retain or lose his interim position as coach of the team. If he performs he becomes the new coach, from being the interim coach. Pedro is correct.

Q: 4 The football competition's final is billed as the most spectacular sporting event of the decade.
There was a massive advertisement in the media with promotional billboards everywhere, and the gate fees were discounted. A full stadium is expected.
But just two hours before the gate was opened a heavy rainstorm started. The roads were flooded and the stadium became waterlogged. Authorities advised everyone to stay at home, and residents of the city agreed.

Barry: "The fans' turnout will be disappointing with this weather."
George: "There will still be a large fan turnout with the extensive promotion for the match."

Which of the following reasoning is most likely correct?

 **A.** Barry Only
B. George Only
C. Both Barry and George
D. Neither Barry nor George

Explanation

Though there were extensive promotions and incentives to make people go to the stadium to watch the match, the bad weather served as a big disincentive for many people to go to the stadium. Option A is right.

Q: 5 Some friends were trying to recall the number of the bus they took from the city centre to the seaside three months ago. They can remember it was a B Bus as it travelled westward out of town to the seaside. The three friends agreed that the bus was an afternoon bus with a 4 and the second to the last number was 7.

If the bus number comprises whether its route is within the city (A) or outside (B). morning (1-), afternoon (2-) or evening (3-) ; and three other numbers, which is the most probable number of the bus the friends took to the seaside?

    A. B1-047

    B. A2-477

    C. A1-741

✓ Correct Ans   D. B2-470

Explanation

The bus number starts with a B as it is an out-of-city bus. Also it is an afternoon bus, so the next number is 2-. With this options A and B and C are wrong. There must be a 4 and 7 in the number with 7 being the second to last number. So Option D is the only logically correct one.

Q: 6 A thief steals a painting from a museum, leaving a cryptic note: "The moon reveals the truth, before darkness sets in." Police discover two hidden doors in the museum, one facing East and one facing West.
Given that:
• The full moon rises in the east.
• Security cameras are activated at sunset.

Through which door did the thief escape?

✓ Correct Ans   A. East door, before cameras activated.

    B. East door, after cameras activated.

    C. West door, before cameras activated.

    D. West door, after cameras activated.

Explanation

The clue refers to the "moon revealing the truth" and "darkness setting in." The full moon rising in the east suggests the thief used the east door for escape. Since it mentions "before darkness," they likely left before sunset and camera activation. So the answer is (a).

Q: 7 A bakery offers muffins in three flavors: blueberry, cranberry, and chocolate chip. Orders include the flavor and additional topping options: nuts, sprinkles, or none. The following are customer requests:
- Order 1: Blueberry with nuts
- Order 2: Cranberry, no topping
- Order 3: Chocolate chip, sprinkles

The bakery accidentally mixes up the toppings. Which order received the wrong topping?

    A. Order 1

    B. Order 2

    C. Order 3

✓ Correct Ans **D.** Impossible to determine

Explanation

We only know what each order requested, not what they received. Without further information, identifying the mistake is impossible.

Q: 8 A city has three bus routes: Red, Green, and Blue; each covers specific areas: North, West, and East, respectively. Given the following information:
- The Red route stops at the City Hall in the North.
- The Green route serves the Museum in the East.
- The Blue route goes past the Train Station in the West.

If a passenger boards a bus at the Library, located in the West, which route should they take to reach close to the Airport in the East?

    A. Red

✓ Correct Ans **B.** Green

    C. Blue

    D. Possible with any route

Explanation

While the Blue route serves the East, it doesn't reach the library in West. The Red route goes to the North (City Hall), not the East. Only the Green route serves both the Library's West location and the Airport in the East (may be close to the Museum at East).

Q: 9 A hotel offers three room types: Standard, Deluxe, and Suite. Each has a different view: city, park, or ocean. Given the following information:
• Standard rooms never have an ocean view.
• Deluxe rooms offer either a city or park view.
• Suites always have an ocean view.

If a guest specifically requests a room with a park view, which type of room should they book?

- A. Standard
- B. Deluxe
- C. Suite
- ✓ Correct Ans D. Impossible to guarantee

Explanation

Standard rooms can have a park view, but not certain; Suites only offer ocean views. This leaves Deluxe rooms as the only option, but it still doesn't guarantee a park view.

Q: 10 A library organizes books by genre (fiction, non-fiction) and publication date (old, new). Each section has specific borrowing limits: 2 books for fiction (old or new), 3 books for non-fiction (old or new).
A student borrows one old fiction book and two new non-fiction books. Did they break any borrowing rules?

- A. Yes, exceeding the fiction limit.
- B. Yes, exceeding the non-fiction limit.
- ✓ Correct Ans C. No, within all borrowing limits.
- D. Information insufficient to determine.

Explanation

Each section has its own limit, and the student didn't exceed any individual limits.

Q: 11 Archaeologists discover a stone tablet with an inscription in an unknown language. They identify recurring symbols and decipher their meanings:
• A circle represents "sun"
• A triangle represents "water"
• A square represents "land"
• A line represents "path".

The inscription reads: "Circle, triangle, line, square, triangle, square, circle." What does the inscription likely convey?

- A. A journey from land to water by a sunlit path.
- ✓ Correct Ans B. A cyclical relationship between sun, water, and land.
- C. A warning about dangers related to sun, water, and land.
- D. Instructions for a ritual involving sun, water, and land.

Explanation

The repetition of symbols and their interconnectedness suggest a cyclical relationship rather than a specific journey or warning.

Q: 12 An explorer finds a faded treasure map with cryptic symbols:
• A skull – "Danger lies ahead."
• A compass pointing north – "Follow the true north."
• A palm tree – "Seek the shade of the tallest."
• A chest – "Your reward awaits beyond the fallen giant."

The explorer finds a skull-shaped rock marking a trail leading north. After following it, they reach a clearing with three palm trees, one significantly taller than the others. Nearby, a fallen tree trunk lies partially buried. Where should the explorer dig?

A. Beneath the skull-shaped rock, assuming it marks the starting point.

B. Under the tallest palm tree, following the "shade" clue literally.

✓ Correct Ans C. At the base of the fallen tree trunk, interpreting "fallen giant" metaphorically.

D. More information is needed about the surrounding area and terrain.

Explanation

Combining the clues suggests a metaphorical interpretation. The path from the skull leads them, not to the "tallest" palm (literal shade), but to the "fallen giant" (tree trunk) where the treasure might be hidden.

Q: 13 Ginseng has been utilized to enhance general health. Additionally, it has been used to boost immunity, aid in illness prevention, and reduce stress. Ginseng of many varieties has been used to treat diabetes, erectile dysfunction in men, and unclear thinking. Ginseng may lower your risk of contracting the flu or a cold. Ginseng users frequently report feeling more alert. Ginseng may enhance performance in tasks requiring mental computation, focus, memory, and other abilities.

If the information is true, which of the following is false?

A. Ginseng enhances and sharpens your memory.

B. Ginseng enhances your brain's function.

✓ Correct Ans C. Ginseng makes you more alert making you restless.

D. Ginseng prevents illness and boosts immunity.

Explanation

The information above says that ginseng may lower your risk of contracting the flu or a cold. Ginseng users frequently report feeling more alert. There is not enough evidence to suggest that ginseng makes you feel more alert, making you restless. Therefore, option C is false.

Q: 14   Which of the following completes the series of figures.

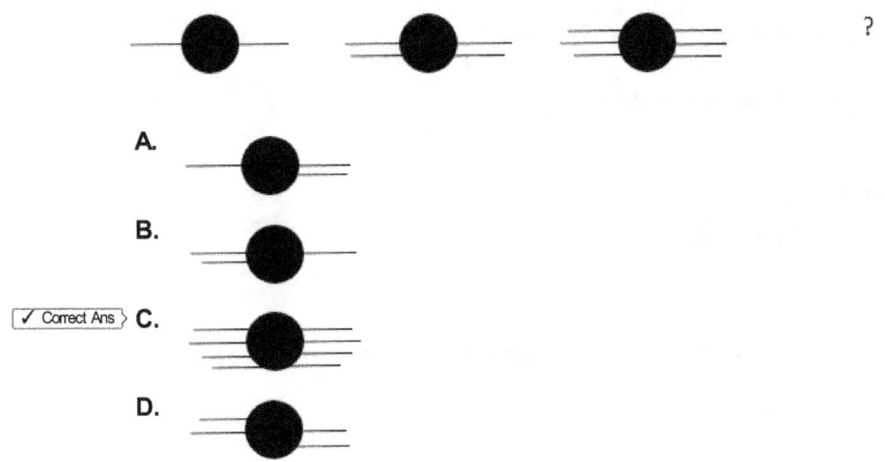

A.
B.
✓ Correct Ans C.
D.

Explanation

The figures above increase the number of lines by 2. Thus if the first figure has 2 lines then the second figure will have 4 lines and third figure will have 6 lines. The last figure will have 8 lines. Option C is the correct answer.

Q: 15   If you start a journey facing north, walk for 6 km, turn right, walk for 2 km, turn left, and walk for 10 km, which direction will you be facing eventually?

  A. West
  B. East
  C. South
✓ Correct Ans D. North

Explanation

The direction won't change; hence, you will also be walking northwards, as shown below.

Q: 16 If you wake up late, you won't have breakfast. If you miss breakfast, you will faint during the school assembly. If you faint, you will miss the morning classes. If you miss the morning classes, you won't see the new teacher.

If Angela missed her morning classes, which of the following is **false**?

- A. Angela woke up early but missed her breakfast
- B. Angela took her breakfast
- ✓ Correct Ans C. Angela saw the new teacher
- D. Angela didn't faint during the assembly

Explanation

If Angela misses morning class, she won't see new teacher. Others choices are not conclusive.

Q: 17 A sports team is trying to qualify for the playoffs. **They have played 30 games so far and won 12 games, and they need to win at least 60% of their games to secure a spot.**

If the information in the bold letters is true, whose reasoning is correct?

Lisa: "If we win 18 of the remaining 20 games, we will secure a spot in the playoffs."
Jake: "To secure a spot in the playoffs, we need to have won at least 48 out of the total 50 games played."

- ✓ Correct Ans A. Lisa only
- B. Jake only
- C. Both Lisa and Jake
- D. Neither Lisa nor Jake

Explanation

If the team has played 30 games so far and needs to win at least 60% of their remaining games to secure a playoff spot, they need to win 18 more games from the remaining 20 games to secure a total of 30 wins. Therefore, Lisa's statement that winning 18 of the remaining 20 games will secure a playoff spot is correct. Jake's statement that they need to have won at least 48 out of the total 50 games played is incorrect because it does not consider the percentage requirement.

Q: 18 Mark argues that using solar energy is the most effective solution to combat climate change. "Solar energy is renewable and produces no greenhouse gas emissions," Mark asserts.

Which one of these statements, if true, most weakens Mark's argument?

✓ Correct Ans **A.** The production and disposal of solar panels can have negative environmental impacts, contributing to pollution.

B. Solar energy is not consistently available in all geographic regions, making it unreliable as the sole energy source.

C. Fossil fuel energy sources are becoming more efficient and cleaner, reducing their environmental impact.

D. Wind energy has been found to be more cost-effective and efficient than solar energy in generating electricity.

Explanation

Mark's argument is based on the belief that using solar energy is the most effective solution to combat climate change. If it is true that the production and disposal of solar panels can have negative environmental impacts, contributing to pollution, it weakens Mark's argument by suggesting that the implementation of solar energy has its own environmental drawbacks. This indicates that the overall environmental impact of solar energy needs to be considered, weakening Mark's argument.

Q: 19 John argues that implementing a four-day workweek for employees is the best solution to improve work-life balance and employee productivity. "A shorter workweek will allow employees more time for personal activities and reduce burnout," John asserts.

Which one of these statements, if true, **most weakens** John's argument?

✓ Correct Ans **A.** Employees may feel more stressed to complete the same amount of work in a shorter period, leading to increased burnout.

B. A shorter workweek has been linked to higher job satisfaction levels among employees.

C. Implementing a four-day workweek may result in reduced pay for employees, leading to financial strain.

D. Other companies that have adopted a four-day workweek have reported increased employee productivity and satisfaction.

Explanation

John's argument is based on the belief that implementing a four-day workweek will improve work-life balance and employee productivity. If it is true that employees may feel more stressed to complete the same amount of work in a shorter period, leading to increased burnout, it weakens John's argument by suggesting that the shorter workweek may have unintended negative consequences on employee well-being. This indicates that the impact of a four-day workweek on employee stress levels needs to be considered, weakening John's argument.

Q: 20 If the product is of high quality, it must have passed rigorous testing. It is mandatory requirement.

If this is true, which one of these sentences must also be true?

- A. If the product has passed rigorous testing, it must be of high quality.
- ✓ Correct Ans B. If the product hasn't passed rigorous testing, it must not be of high quality.
- C. If the product has passed rigorous testing, it cannot be of high quality.
- D. If the product hasn't passed rigorous testing, it must be of high quality.

Explanation

Rigorous testing ensures high quality.

Q: 21 During the Christmas period a store offers a special promotion on its household products on sale with discounted prices for the products. This was based on a code attached to the price tag. Shoppers are expected to add the two numbers attached to the price tag and subtract their sum from the original price to find the selling promo price of the items. For example, a $20 item has "23", attached, implying that the discount percentage is $5/20 = 25\%$

From the following 3 items, which one offers the most percentage of discounts to shoppers?
- Item A: $32, "35"
- Item B: $20, "42"
- Item C: $48, "75"

- A. Item A
- ✓ Correct Ans B. Item B
- C. Item C
- D. All offer the same discounts

Explanation

The discounts for each item:
A: 3 + 5 = $8 ; $8/35$ = 22.85 %
B : 4 + 2 = $6 ; $6/20$ = 30 %
C: 7 + 5 = $12; $12/48$ = 25%
So B offers the most discount; it is the correct option.

Q: 22 Tessy is looking for a suitable location to open her new gift store specializing in children's birthday gifts. She is looking for a shop not too far from the main street in town and one that would attract the right kind of foot traffic.

Which of the following locations shown to her by the realtor is appropriate for her needs?

I. A store in a small mall located just on the outskirts of town, with two other gift stores and a saloon in the mall.
II. A big store on the ground floor of a new high-rise building about 20 minutes from the centre of town.
III. A little shop about two blocks away from the town's main street, located across the street from a popular small grocery mall and beside a female saloon.
IV. A stand-alone big store on a quiet residential street 20 minutes away from the town's centre

- A. I Only
- B. II Only
- ✓ Correct Ans C. III Only
- D. IV Only

Explanation

This best option is III. It is near the main street, just two blocks away, and a location near a popular option is both near a popular grocery mall and a female saloon. These two commercial outlets will guarantee good foot traffic for the gift store. Option C is correct.

Q: 23 Your cousin just sent you a message. He is sending a gift to your father for his 60th birthday. It is a special gift.
"It has a face and two hands but no arms or legs. But its hands cannot clap."

What can be the gift?

- A. A shirt
- B. A painting
- ✓ Correct Ans C. A clock
- D. A smartphone

Explanation

A clock has a face, and two hands: minutes and hours but has no arms or legs. Its hands cannot clap with each other. Therefore, option (C) is the correct answer to the riddle.

Q: 24 The lock to a box is opened by a set of 4-alphabetic letters given by the following clues:
- the first letter starts the car
- The second letter ends the whole chant
- The third starts the whole chant
- The fourth starts the race
- And the whole is the emperor.

Which set of letters is the code that opens the lock?

- A. K, E, Y, S
- ✓ Correct Ans B. C, Z, A, R
- C. B, E, S, T
- D. D, A, R, T

Explanation

Analyze the clues one by one;
- the first letter starts the car: C
- The second letter ends the whole chant: Z
- The third starts the whole chant: A
- The fourth starts the race: R
- And the whole is the emperor: CZAR
The correct option is B.

Q: 25 A newspaper publisher assigns unique identification numbers to its newspaper using the following system:
- The first two digits represent the year it is published
- The third and fourth represent the quarter
- The fifth and sixth represent month
- The last two digits represent the day of the month published

If a newspaper's ID number is 82020523, which of the following statements must be false?

- A. It may be published in 1982.
- ✓ Correct Ans B. It was published in June
- C. It was published in the second quarter
- D. It was published on the 23rd of a month

Explanation

The first two digits are 82 – indicate it was published in 1982 or 1882
The third and fourth digits – 02 indicate it was published in the second quarter
The fifth and sixth digits - 05 indicate it was published in the fifth month – May
The seventh and eighth digits show it was published on the 23rd of a month.
All other options are true, except Option B.

Q: 26 There is a serial burglar on the loose on the block. The Policemen investigating the series of break-ins got the description of the suspect from 4 residents, who claimed to have seen the suspect leaving some of the apartments that were broken into.
-Witness 1: The suspect is a Latino male, in his mid-twenties, 5'3", with short dark hair and a broad chest. He was wearing dark blue jeans and a blue T-shirt.
-Witness 2: The suspect is a Mexican Male, aged 45-50 years old, 5'10", 245 pounds, with a shaved head. He was wearing a navy blue blazer and brown trousers.
-Witness 3: The suspect is dark-skinned, male, approximately 25 years old, 5'9", slight build, with short dark hair. He was wearing a blue suit.

During a Police surveillance of the block, a suspect was caught breaking into an apartment.

Description of suspect: He is a dark-skinned male, aged 24 years old, 140 pounds, about 5'9", with short dark hair. He was wearing a dark blazer and a pair of brown trousers.

Based on the descriptions by the three earlier witnesses, which of these witness descriptions was most likely that of the caught suspect?

A. Witness 1

B. Witness 2

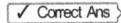 C. Witness 3

D. Any of them

Explanation

Witness 1 described the suspect as 5'3", with short brown hair and broad-chested, these are at variance to the caught suspect's height, hair and build.
Witness 2 described the suspect as having a shaved head heavily built and being 5' 10" in height. Somewhat like the caught suspect.
Witness 3 described the suspect as a dark-skinned, male, approximately 24 years old, 5'9", slight build, with short dark hair, which is the closest description to the caught suspect.
Option C is correct.

Q: 27 A restaurant offers three types of burgers: classic, bacon cheeseburger, and veggie burger. The menu states:
• All burgers come with fries.
• Bacon cheeseburgers do not include a drink.
• Veggie burgers come with either a salad or onion rings.

Which option is impossible?

A. Ordering a classic burger with a salad.

B. Ordering a bacon cheeseburger and a drink.

C. Ordering a veggie burger with fries and onion rings.

 D. Ordering a classic burger without fries.

Explanation

Every burger comes with fries. Removing fries for the classic burger violates this rule.

Q: 28 Three trains, the Comet, the Express, and the Local, travel between City A and City B. Their travel times are unique:
• The Comet takes 4 hours.
• The Express takes 5 hours.
• The Local takes 8 hours.

If two trains leave City A simultaneously heading for City B and arrive at their destination with a 3-hour difference, which trains were they?

A. Comet and Express

B. Comet and Local

✓ Correct Ans C. Express and Local

D. Information insufficient

Explanation

Only a 3-hour difference exists between the Express and Local (5 hours vs. 8 hours), which matches the scenario.

Q: 29 Four friends, Alice, Bob, Charlie, and Damien, live in a building with four floors. Each friend lives on a different floor, and nobody shares a floor. They gave clues about their residences:
• Alice lives neither on the top nor the bottom floor.
• Bob doesn't live on the ground floor.
• Charlie is two floors above Damien.

Who lives on which floor?

✓ Correct Ans A. Alice – 2nd, Bob – 4th, Charlie – 3rd, Damien – 1st

B. Alice – 3rd, Bob – 2nd, Charlie – 1st, Damien – 4th

C. Alice – 2nd, Bob – 1st, Charlie – 4th, Damien – 3rd

D. Alice – 3rd, Bob – 4th, Charlie – 2nd, Damien – 1st

Explanation

- Alice can't be on the top or bottom floor, leaving floors 2 and 3.
- Bob isn't on the ground floor, eliminating the 1st floor and leaving the 2nd, 3rd, and 4th floors.
- Charlie is two floors above Damien, so he can be 3rd or 4th floor.

So, only Damien can take 1st. If Damien is 1st, Charlie is 3rd. Alice now can only take 2nd, leaving 4th to Bob.

Q: 30   A group of friends organized a potluck party. Each person agreed to bring a dish that starts with a specific letter of the alphabet, ensuring a diverse spread. Here are their assignments:
• Ann: Appetizing salad
• Boy: Bread for lunch
• Claire: Casserole
• Donny: Desserts like apple pie
• Emma: Entrée like lasagna

However, on the day of the party, some changes occurred:
Ann brought a salad.
Boy brought sweet dessert instead of bread.
Claire couldn't find a casserole recipe and brought chips.
Donny baked an apple pie.
Emma, true to her assignment, brought a delicious lasagna.

Which dish does not fulfil either the alphabetical assignment or the food type assignment?

    A. Salad

 **B.** Sweet dessert

    C. Chips and dip

    D. Apple pie

Explanation

Analyze each dish considering the original assignments and their starting letters:
• Salad (Ann) – Fulfills "Appetizer" assignment but not the letter "A".
• Sweet dessert (Boy) – Violates "Bread" assignment and doesn't start with "B".
• Chips (Claire) – Deviates from "Casserole" but does start with 'C'.
• Apple pie (Donny) – Completes "Dessert" assignment.
• Lasagna (Emma) – Perfectly aligns with lasagna.

Therefore, sweet dessert (option B) is the only dish that doesn't fulfill either the food type assignment or the starting letter requirement.

Q: 31   A health scientist tells that:

"*Even though chemical fertilizers and pesticides are some of the main developments in the agricultural sector, they are a major cause of water pollution, and they seriously affect our kidneys too.*"

Which of these statements, if true, best supports the scientist's claim?

    A. Organic fertilizers and natural pesticides are way better and more effective than chemical fertilizers and pesticides.

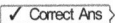 **B.** When people drink that polluted water, the chemicals in chemical fertilizers and pesticides impair their kidneys.

    C. When walking on agricultural fields, chemical fertilizers and pesticides may go through the wounds on the skin, which may cause kidney failures.

    D. Chemical fertilizers and pesticides can be recycled.

Explanation

Scientist's claim is chemical fertilizers and pesticides are a major cause of water pollution, and they seriously affect our kidneys too. So the supporting statement should explain how water pollution affects human kidneys.

Q: 32 If James improves his coding skills, he's likely to become a more efficient programmer. Becoming a more efficient programmer will help him complete projects faster. The only other way James can complete projects faster is by attending coding workshops.

Which one of the following is not possible?

A. James improved his coding skills and became a more efficient programmer, which helped him complete projects faster.

B. James did not improve his coding skills, but he attended coding workshops and started completing projects faster.

C. James did not improve his coding skills, did not attend coding workshops, and could not complete projects faster.

✓ Correct Ans D. All are possible

Explanation

Without improving coding skills, completing project faster is still possible by attending coding workshops. All are possibilities.

Q: 33 Mike: "To become proficient in a foreign language, you should practice speaking with native speakers."
Lisa: "I had a conversation with a native speaker last week, so I'm fluent now."

Which sentence best points out the incorrect **assumption behind** Lisa's argument?

A. Regular conversations with native speakers enhance language skills.

B. Lisa's fluency in the language has nothing to do with speaking with native speakers.

✓ Correct Ans C. Speaking occasionally with native speakers guarantees language proficiency.

D. Mike's advice about practising with native speakers is inaccurate.

Explanation

Lisa mistakenly believes that a single conversation with a native speaker is enough to achieve fluency, whereas the recommendation implies consistent practice over time.

Q: 34 "To excel in basketball, one needs to have height, agility and good coordination."

Mark: "Sarah is tall and has excellent coordination. She's destined to excel in basketball."
Lily: "However, Sarah lacks the agility and quick reflexes that are also crucial for basketball success. She can't excel in basketball"

Whose reasoning is correct?

A. Mark only

✓ Correct Ans B. Lily only

C. Both Mark and Lily

D. Neither Mark nor Lily

Explanation

Agility is a mandatory attribute to excel in basketball.

Q: 35 *If a laptop is a gaming laptop, it is expensive.*
George bought an expensive laptop yesterday.

What is the conclusion about George's laptop ?

   A. It is a gaming laptop

   B. It isn't a gaming laptop

   C. It is expensive but not a gaming laptop

✓ Correct Ans **D.** No conclusion

Explanation

If a laptop is a gaming laptop, then it's expensive. But that statement doesn't work in the opposite direction. In other words, we cannot tell that if a laptop is expensive then it's a gaming laptop. So we cannot arrive at any conclusion about George's laptop.

Q: 36 There are three special cards in a deck: "Star" "Moon," and "Sun."
The following rules apply:
- The "Star" is superior if and only if "Moon" is not on display.
- "Sun" is superior even if "Moon" is displayed.
- "Moon" is superior only if the "Sun" is hidden.

Which of the following statements must be true?

   A. The least valuable card is "Moon"

   B. All three cards are superior.

   C. Only "Star" is a valuable card.

✓ Correct Ans **D.** "Sun" is the most valuable card.

Explanation

The "Star" is superior only if the "Moon" is not displayed, so "Moon" is more valuable than "Star"
"Sun" is superior even if "Moon" is displayed, so the "Sun" is more valuable than "Moon".
So the "Sun" is the most valuable of the three cards. Option D is right.

Q: 37 In a display of various polygons and numbers, the reveals were in the following order:
- hexagon (12)
- pentagon (10)
- rectangle (8)

What will be the logical next shape and accompanying figure?

   A. Octagon (14)

   B. Square (4)

✓ Correct Ans C. Triangle (6)

   D. Heptagon (3)

Explanation

In the reveal pattern, the number of the polygon's sides decreases by one each time just as the figure in the circle increases by 2.
The first polygon is a hexagon (6 sides) with the number 12; the next is a pentagon – 5 sides with 10; the next rectangle (4 sides) with 8, so the next will be 3 sides (triangle with the number 6. Option c is right.

Q: 38 There are special cards that can be drawn in a game:
- The Blue card gives you an addition of 5 marks;
- The Red card deducts 5 marks from your total;
- The Yellow card can alternate with an addition or subtraction of 5 marks each time you draw it, with a subtraction first.

Based on these rules, what will be your marks addition or subtraction if you draw a Yellow card, a Red card, a Red card, and a Blue card in that order?

   A. +10

✓ Correct Ans B. -10

   C. +15

   D. -15

Explanation

The cards are drawn in this order:
Yellow card (-5), Red card (-5), Red card (-5), and a blue card (+5) = -10
Option B is correct.

Q: 39 Four friends, Mia, Noah, Olivia, and Peter, are discussing their favorite movie genres:
- Mia loves comedies and animation.
- Noah is a big fan of dramatic action and thrilling sci-fi movies.
- Olivia prefers dramas and documentaries on current affairs.
- Peter enjoys action-packed thrillers and mysteries.

There's a new movie festival with four categories: Action, Comedy, Documentary, and Thriller. Each friend can only choose one movie to watch.
Knowing their preferences, which movie category is least likely to be chosen by any of them?

- A. Action
- B. Comedy
- ✓ Correct Ans C. Documentary
- D. Thriller

Explanation

* While not everyone's favorite, both Noah (action/sci-fi) and Peter (thrillers/mysteries) might be interested in action movies.
* Mia loves comedies, and Olivia might enjoy a documentary depending on the subject.
* Peter loves thrillers and Noah might enjoy that too.
* Documentaries are the least likely choice because it's a possibility for Olivia only, but depends on subject.

Q: 40 Identify the wrong term.
14, 41, 52, 25, 36, 63, 74, 47, 56, 85

- A. 74
- B. 47
- C. 56
- ✓ Correct Ans D. 85

Explanation

14 → 41, 52 → 25, 36 → 63,
74 → 47, 56 → 85 (65)
Each number has been interchanged.

www.ingramcontent.com/pod-product-compliance
Lightning Source LLC
Chambersburg PA
CBHW080036100526
44584CB00023BA/3224